武汉纺织大学教材出版基金资助出版

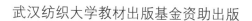

纺织英语

主　编　郑亚娟　曾令宙　杨　雪
副主编　李建鹏　李燕芬　刘方荣　徐凌琳
参　编　刘　珊　王　辞

武汉大学出版社

图书在版编目(CIP)数据

纺织英语/郑亚娟,曾令宙,杨雪主编.—武汉:武汉大学出版社,2021.5
ISBN 978-7-307-22288-5

Ⅰ.纺… Ⅱ.①郑… ②曾… ③杨… Ⅲ.纺织—英语 Ⅳ.TS1

中国版本图书馆 CIP 数据核字(2021)第 092604 号

责任编辑:罗晓华　　　责任校对:李孟潇　　　版式设计:韩闻锦

出版发行:**武汉大学出版社**　(430072　武昌　珞珈山)
　　　　　(电子邮箱:cbs22@whu.edu.cn　网址:www.wdp.com.cn)
印刷:湖北金海印务有限公司
开本:787×1092　1/16　印张:16　字数:367 千字　插页:1
版次:2021 年 5 月第 1 版　　2021 年 5 月第 1 次印刷
ISBN 978-7-307-22288-5　　定价:42.00 元

版权所有,不得翻印;凡购我社的图书,如有质量问题,请与当地图书销售部门联系调换。

前　言

《纺织英语》教材基于课堂教学实践并吸收国内外同类教材优点编写而成。在"新文科"建设大背景下，本书的编写以"语言产出能力"为导向，采取"语块为中心的数据驱动研究型学习"教学理念，在目标词汇的遴选方式、编排逻辑、呈现方式、练习设计中充分利用语料库，凸显专业语块的核心地位，如创设语料库数据驱动式检索猜词任务，引导学生通过多个语境推理和猜测目标语块，促进语言数据挖掘、数据分析能力的提升，培养社会科学家。

本教材既重视语言学习，又带有纺织科普性质。

一、教材理念

1. 学生中心。本教材充分考虑学生的语言基础及对纺织服装主题的已有知识经验，对语料进行了删节、加工，阅读难度适中；在章节内容选择上，充分考虑学生的兴趣及科普需要，既关注历史，也有对现实及未来纺织故事的讲述，既有远方的、科技的、产业的、经济的纺织话题，也有身边的、生活的纺织知识科普；在呈现方式上，尽量以图文并茂的方式激发学生的探究兴趣。

2. 产出导向。本教材内容丰富，有机结合了英语语言技能训练与纺织知识掌握，既关注学生语言水平的提高，也聚焦纺织相关知识的传授。教材编写既注重内部系统性，也关注使用的外部有效性。材料和讲授内容充分考虑了专业基础性、实用性及趣味性，以满足高等学校学生和相关人士了解纺织及做好纺织涉外工作的需要。

二、教材内容

本教材共分六章，语料涵盖纺织故事、历史、文化、生产、经济及应用。语言训练包括阅读、语块应用、口头表达、书面表达，并配以相应练习。每一章最后还规定了探究性思考话题，以进一步培养学习者的自学能力。

第一章　纺织导论

本章分为两节。第一节简要回顾古代纺织技术发展和相关历史文物，重点介绍中国的丝绸工艺和提花织机的发展及其对文化艺术和生产贸易的影响。第二节主要介绍世界知名纺织院校，包括英国利兹大学、曼彻斯特大学，美国北卡罗来纳州立大学，德国亚琛工业大学，俄罗斯莫斯科国立纺织大学，印度理工学院德里分校，日本信州大学，中国武汉纺织大学。

第二章　纺织历史与文化

本章分为两节。第一节从中国纺织历史角度，重点介绍丝绸的历史和棉花的历史，以及中国纺织历史代表人物：嫘祖、黄道婆。第二节主要介绍纺织与中国传统艺术文化的联系，包括纺织业在汉赋中的呈现，纺织与中国传统艺术形式，如四大织锦、著名刺绣等。

第三章　纺织与生活

本章分为两节。第一节介绍服饰的内涵，服饰穿着的理由，服饰的历史以及不同时期服饰时尚的特点。第二节主要从纺织的角度，从质地、图案、颜色和织物四个方面阐述纺织与服饰的关系。另外搭配补充材料，从流行元素探索服饰的发展与变迁。

第四章　纺织与生产

本章分为五节。第一节从天然纤维和人造纤维两个方面来介绍纺织纤维的种类和纤维的性质。第二节介绍纱线的加工和制作过程，主要对人造纤维的纺丝和天然纤维的纺纱两种方法进行讲解。第三节介绍机织的基本原理和织造过程，机织物的基本组织和性质。第四节从纬编和经编两种编织方式来介绍针织的基本原理和成圈过程，针织物的基本组织和性质。第五节介绍了非织造织物常见的七种生产过程。

第五章　纺织与经济

纺织业从古至今一直是国民经济中的一个重要部分。本章分为三节。第一节借助古代丝绸之路介绍纺织贸易。从纺织品出现开始就伴随着纺织贸易，丝绸之路（或许更早）是纺织品国际贸易的典型代表，凸显了纺织品在经济中的作用。第二节介绍在疫情下，纺织工业对经济恢复的可能帮助。第三节介绍纺织品的循环利用与绿色经济的关系。人类社会发展至今愈发重视环境保护和绿色发展。纺织品在生产、使用过程中存在大量污染环境和浪费资源现象。如果全行业和社会能重视这个问题，采取积极措施，就能减少浪费，为绿色发展作贡献。本章另外附加阅读材料，预测纺织品行业的未来贸易趋势。

第六章　技术纺织品与智能纺织品

本章分为三节。第一节从整体上概述了技术纺织品的 12 个应用领域，让读者对于技术纺织品的应用有一个完整的框架性认知。第二节重点从防护性纺织品、医用纺织品和汽车用纺织品三个方面介绍了读者相对较为熟悉的几个技术纺织品运用实例。第三节重点介绍了智能纺织品的开发和应用。

本书由武汉纺织大学外国语学院教师编写，各章节编写人员如下：第一章曾令宙；第二章杨雪；第三章刘方荣、徐凌琳；第四章郑亚娟；第五章李建鹏；第六章李燕芬、曾令宙；参编人员刘珊、王辞。在此，对编写组全体成员的通力合作表示诚挚的感谢，同时感谢武汉纺织大学纺织学院李建强教授、吴济宏教授、凌文漪教授的大力支持和帮助！此外，本书在编写过程中参考了诸多学者的著作和相关网站的资料，在此一并表示敬意和谢意！由于时间仓促、编者水平有限，难免出现疏漏和错误，恳请广大读者批评指正。

谭燕保
2021 年 3 月

Contents

Chapter 1　The Introduction of Textile 1
   Section 1　Heritage 1
   Section 2　Transmission 15

Chapter 2　Textile History and Culture 32
   Section 1　Textile History in Ancient China 32
   Section 2　Textile and Culture 45

Chapter 3　Textile and Life 64
   Section 1　About the Clothing 64
   Section 2　Textiles and Clothing 72

Chapter 4　Textile and Manufacture 90
   Section 1　Textile Fibres 90
   Section 2　Yarns 105
   Section 3　Weaving and Woven Fabrics 119
   Section 4　Knitting and Knitted Fabrics 129
   Section 5　Nonwoven Fabrics 145

Chapter 5　Textile and Economy 156
   Section 1　Textile and Trade 156
   Section 2　The Textile Industry and Economic Recovery 162
   Section 3　Recyclable Textiles for Green Economy 167

Chapter 6　Technical Textiles and Smart Fabrics 180
   Section 1　General Introduction to Technical Textiles 180
   Section 2　Detailed Introduction to Some Technical Textiles 191
   Section 3　Smart Fabrics 205

Contents

Glossary 212

References 247

Chapter 1　The Introduction of Textile

Section 1　Heritage

Lead-in

Warm-up questions:

What is the relation between computer programming and textile weaving?

Lexical chunk bank	
植物纤维	vegetable fibre, plant fibre
锭盘	spindle whorl
韧皮纤维；麻纤维	bast fibre
纺车	spindle wheel, spinning wheel
平纹组织	plain weave
刺绣图案	embroidered design
底布	ground material, ground fabric
对称组织	symmetrical organization
缠枝纹	twining foliage
优美的曲线	sinuous curves
纱罗织物	gauze weave
绒圈锦	pile-loop brocade
经丝显花	warp patterning

卧式织机；踞织机	horizontal loom, ground loom
立式织机；竖机	vertical loom, upright loom
经轴	warp beam
踏式织机；蹑机	treadle loom
手工提花织机；花机	drawloom
挽花工	drawboy
纬丝显花	weft patterning
防染；扎染；夹缬	resist dyeing
室内装饰面料	upholstery fabric
斜纹锦缎	twill damask
提花织机	pattern loom
图案单元；花本	pattern unit

The term "textile" is a Latin word originated from the word "*texere*" which means "*to weave*". Textile refers to a flexible material comprising of a network of natural or artificial fibres, known as **yarn**. Textiles are formed by weaving, knitting, **crocheting**, knotting, and pressing fibres together.

As a major component of material culture, textiles may be viewed as the products of technology, as cultural symbols, as works of art, and as items of trade.

Humans have made and worn textiles since prehistoric times. Archaeologists and anthropologists suppose that prehistoric peoples produced and wove clothing based on evidence from archaeological digs. They have uncovered bone fragments that could have been used as needles, statues depicting clothed figures, and fragments of fibres that could have been cloth.

The research of textile history depends on the chance survival through the centuries of materials inherently prone to decay. Textiles are made to be used primarily as **furnishings** and dress, and are expected to wear out and eventually be discarded. It is for this reason that a study like this is essentially histories of decorated textiles, since the ordinary or the everyday will rarely survive the lifetime of its user.

In many societies textiles have played a vital role in the social, economic and religious life of the community. They are essential **accoutrements** in all major life-cycle ceremonies such as births, weddings, and funerals, when they are either bestowed as gifts, exchanged, burned, buried, or passed on to the next generation as the substance of dynasty.

In comparison with objects of metal, stone, pottery or glass, textiles represent only a

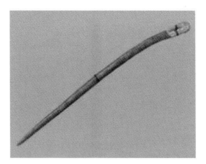

Bone needle found in the Peking Man site at Zhoukoudian, Beijing, c. 18,000-11,000 BC

Jade sculpture of a seated human figure from Fuhao's tomb, Henan Province, c. 1250-1192 BC

small proportion of excavated material. In damp soil the **vegetable fibres** of which many textiles are made disintegrate entirely, although occasionally the copper salts from oxidized bronze or silver objects pinned to articles of clothing will act to preserve small fragments of cloth from decay. Nevertheless, special circumstances in several different parts of the world have meant that some textiles have been preserved over extraordinarily long periods of time.

Flax was the commonest ancient plant fibre, though **hemp**, **rush**, palm, and **papyrus** were also used. Probably originating in West Asia, this plant requires a good water-supply and settled, organized communities for its cultivation. Seeds of domesticated flax (*Linum usitatissimum*) found with **spindle whorls** together on the same site, are indicative of textile activity.

Chinese textile history largely focuses on silk. Although other **fabrics** do not now survive in such quantities as silk, three types of **bast-fibre** material were in use from Neolithic times up to the thirteenth century. These were **ramie**, hemp, and *ge* from the fibres of *Pueraria thunbergia*, the vine-creeper. Cotton, an import from India, grew in economic importance during the fourteenth century and remained the everyday material of the Chinese people.

The Chinese invention of the **spindle wheel** during the Zhou Dynasty (1046-256 BC)

Chapter 1   The Introduction of Textile

T-shaped painting on silk from Xin Zhui's tomb, Hunan Province, c. 168 BC

Bundles of vegetable fibres from ancient Egypt, now in the Egyptian Museum, Turin, Italy

was first applied to bast fibres and then adapted for silk. Silk is lightweight, strong, and naturally beautiful. It can be woven, spun, dyed, and worked in a variety of ways. These properties make it a **lustrous** fabric, primarily produced for the prosperous sector of society.

The earliest woven silk fragments found to date come from the period between 2860 and 2650 BC. Although these examples were in **plain weave**, it is supposed that fabrics with patterns followed on soon after.

The tomb of an unknown lady provides us with truly astonishing proof of the early range of textile techniques and the intricate patterns that were employed. Dating to about 300 BC, the excavated chamber at Mashan in what is today Hubei Province in south central China was, at the time of the woman's funeral, in the powerful state of Chu. This state, known for its **evocative** poetry and its **singular** decoration on bronze vessels, must now rank high in any world view of the textile arts.

The excavations revealed garments and **bordered shrouds** as well as other smaller textile pieces, and the find is arguably the largest and earliest group of intact textile items ever recovered. Except for a miniature ritual robe, all the **gowns** found were on the woman's corpse, and these, together with the wraps that also enveloped her body, are decorated mostly with complex **embroidered designs** in rows of close chain stitch.

The **ground material** of these is of even, plain-weave silk, and the embroidery spreads across the surface on the diagonal. On some of the robes the pattern repeats are long and, although the **symmetrical organization** of the designs is immediately apparent, the axes about which the patterns turn are often hard to pick out. The **twining foliage** and fantastic birds and beasts dissolve into each other in **sinuous curves**, giving the impression that the design scheme is freer than it actually is.

Detail (LEFT) of a woman's embroidered shroud (RIGHT) from Mashan, Hubei Province, c. 300 BC

Tombs of the Han Dynasty (202 BC-220 AD) have yielded a variety of silks, including plain weave, **gauze weave**, both plain and patterned, and **pile-loop brocade** similar to velvet. The long history of **warp patterning** in China was due to the availability of strong silk thread which, because it did not need laborious spinning, could be used for closely set warps.

The loom is the device for keeping the warp threads evenly spaced and under tension. Really very complex work can be produced on the simplest of looms, and, broadly speaking, developments in loom design have served the dual purpose of making certain processes easier and of increasing the speed with which they could be accomplished. The earliest looms of which we have knowledge are those of the Egyptians: a **horizontal** or **ground loom** (where the warps lie parallel with the ground) is depicted on Egyptian pottery of the Pre-Dynastic Period (*c.* 5000-3100 BC), and a **vertical** or **upright loom** is shown in tomb paintings from the Twelfth Dynasty onwards (*c.* 1900 BC).

On the looms described above the **sheds** are formed by hand, which can involve considerable delay between the passing of each **weft**. If, however, the loom is fitted with a **treadle**, the shed is formed by the action of the feet, leaving the hands free to pass the shuttle with greater speed. It also allows far longer pieces of cloth to be woven, since a very long warp can be wound up on the **warp beam**.

Later Han silks include a striking number of woven patterns with texts, usually several characters with **auspicious** meanings. From pictorial representations, scholars have deduced that Han weavers used **treadle looms**.

Brocade armguard with ideographs woven into the patterns from the Niya entombment, Xinjiang Province, 265-317AD. The inscription "五星出东方利中国" (Wuxing chu dongfang li Zhongguo) means when the Five Planets rise in the East, the advantage will be to China.

The period between the Han and Tang Dynasties—the third to sixth centuries—witnessed a changing aesthetic in textile design as well as a major change in loom arrangement for patterned silk manufacture.

Finds in remote areas have added to our understanding of production and commerce relating to silk textiles. The Silk Road, in reality several East-West routes across the north of

China and beyond, continued the exchange goods and influences between East and West Asia and, ultimately, Europe. Buddhism spread to China from India this way, and a dedicatory textile found at *Mogaoku* in Gansu Province in the Silk Road region represents the first securely dated silk connected with this great faith. Fragments of silk at Noin-Ula, in northern Mongolia, dated second century BC, give further evidence of the widespread exchange of silks throughout Asia. Although details of the trade are yet to be fully understood, comments by early writers make clear the admiration for Chinese silks in the Roman world.

Chinese archaeological exploration has provided an abundance of whole silk **artefacts** the like of which do not exist elsewhere. During the Tang Dynasty (618-907), the **drawloom** was introduced to or independently developed by the Chinese. It has been said that every advanced weaving technique known in the world at the time was practiced by the Tang weavers.

The treadle loom could be modified to produce more complex weaves and to speed production by increasing the number of foot-operated heddles, but there remains a limit to the number of heddles which can conveniently be operated by foot treadles alone.

The drawloom enabled an infinitely greater variety of sheds to be formed. Instead of heddles, the warps on the drawloom which were to be raised for each shed were attached by strings to a cord which passed over the loom. Each shed, therefore, had a cord which was numbered and which was pulled in sequence by a "**drawboy**", while the weaver inserted the weft. As many different sheds were possible as there were combinations of draw cords. The drawboy first sat on top of the loom; later he could stand on the floor and pull down the cords.

This meant that multicoloured patterns could now be created by the weft, giving smoother design outlines and greater sophistication in patterning. Despite the advantages of weft patterning, the changeover from warp to weft ornamentation was a gradual process, and both methods existed alongside each other in the centuries before the Tang Dynasty and on into the early and middle years of that dynasty.

A major departure from loom patterning was the production of refined **resist-dyed** textiles. This involved tying off or masking out with wax or rice paste areas of pattern before immersion of the textile in the dye. A process whereby selective areas of the silk are **degummed** so that they accept the dye more readily than the parts where the **gum** remains seems also to have been employed, as does a method which used paired wooden boards, identically carved in **relief**. Folded silk was laid between these two boards and dye poured through holes drilled in the top one. The color penetrated the silk only at the points where it was not clamped tightly.

For the Post-Tang Era silks which have survived above ground as opposed to those from excavations have to be taken account of in any overview of Chinese textiles. Generally, their numbers increase the closer they are to the present. The Song Dynasty (960-1279) is always

Chapter 1　The Introduction of Textile

Reproduction of indigo resist dyeing

thought of as the era when silk **tapestry** weave, often referred to by its Chinese name *kesi*, or *k'o-ssu*, reached unprecedented heights. The *kesi* pieces which confirm this have been held by collectors down the centuries and preserved as works of art.

Lotus Pond and Ducks, silk tapestry with kesi weaving by Zhu Kerou of the Southern Song Dynasty (1127-1279)

Birds and flowers are perhaps the most recurring **motifs** found on *kesi* pieces of this time, but these textiles belong firmly in the Chinese painting tradition and are of a very different order from the textiles recovered from tombs. Despite being woven, they resemble paintings and some of them are in fact copies of famous paintings. The picture-making opportunities of the *kesi* technique were utilized to the full to produce these valued scrolls and album leaves. The method is akin to that used for the great wool tapestries of Europe.

The Ming Dynasty (1368-1644) and the succeeding Qing Dynasty (1616-1911) had flourishing textile industries in a number of places throughout the Empire. Study of Ming and Qing textiles has focused on the court collections, including wall decorations, curtains, desk **frontals** and **upholstery fabrics**, ceremonial and informal costumes, and works of art.

Square pictorial badges displayed on the front and back of the surcoat of the Qing Dynasty (1616-1911), which was a required item of dress in the Chinese imperial wardrobe, identified the various official ranks at the imperial court. The golden pheasant denotes a second-rank civil servant.

No single technique or group of textiles is more noteworthy than the rest two centuries for there is a mass of **extant** material, all equally interesting. The so-called fine-art tradition continued with the production of *kesi* pictures, but increasingly this weave was applied to garments and furnishings as well.

The drawloom was developed in the Middle East for silk weaving during the sixth and seventh centuries AD, but its origins may lie even further back. Fragments of **twill damasks** dated to the second century BC in China suggest that some sort of **pattern loom** was known there by that time. It remained in use until replaced by the Jacquard loom in the early nineteenth century and was used to produce some of the most elaborate woven textiles ever known. The size and complexity of the drawloom demanded skilled workers and permanent

Chapter 1 The Introduction of Textile

Drawloom in the Drawings of Tilling and Weaving (1979 reprint of 1696 edition). More complex weaves required a two-person drawloom. The person perched above moves the heddles that allow the weaving of complex multi-colored patterns.

positioning of the loom; thus commercial production of patterned textiles became not only possible but, indeed, necessary because of the time required to set up the loom.

During the Industrial Revolution of the eighteenth and nineteenth centuries developing technology allowed the shuttle to be propelled mechanically and sheds to be selected automatically. The Jacquard loom, which was in use in English cotton mills by 1813 and which was adapted for fully mechanized operation during the 1830s, replaced the cords and drawboy by a mechanism driven by a single treadle and which could be operated by the weaver unaided. A series of linked, punched cards, with each punched hole corresponding to one warp end in a **pattern unit**, is mechanically matched to the heddles. When the card mechanism has completed a **revolution**, the different sheds have been passed and the pattern unit is complete.

During the same period water and steam were used to power the looms. However, the basic techniques of weaving remained unchanged. Industrialization brought about greater speed of production and a reduction in the skilled labor required to operate the looms, rather than any improvement in the design or structure of the fabrics produced.

The introduction of a capitalist production system in so many parts of the world during the past two centuries has radically reordered the role of textiles in society. Jane Schneider and Annette Weiner, in their anthology *Cloth and Human Experience*, suggest that, firstly, it eliminates the opportunity for textile producers (spinners, weavers, dyers, and so on) to infuse what they produce with spiritual value. Secondly, in requiring endless variation and rapid **turnover** capitalism has encouraged the growth of the modern fashion system, so that dress has now become the predominant means of expression.

Words:

yarn [jɑːn]	n. 纱线
crocheting [krəʊˈʃeɪɪŋ]	n. 钩针编织
furnishings [ˈfɜː(r)nɪʃɪŋz]	n. 家具陈设
accoutrements [əˈkuːtrəmənts]	n.〈正式〉装备；配备
flax [flæks]	n. 亚麻；亚麻纤维
hemp [hemp]	n. 大麻
rush [rʌʃ]	n. 灯芯草
papyrus [pəˈpaɪrəs]	n. 纸莎草
spindle [ˈspɪnd(ə)l]	n. 轴；纺锤；纱锭
fabric [ˈfæbrɪk]	n. 织物；布料
ramie [ˈræmɪ]	n. 苎麻；苎麻纤维
lustrous [ˈlʌstrəs]	adj. 柔软光亮的
evocative [ɪˈvɒkətɪv]	adj. 引起记忆的；唤起感情的
singular [ˈsɪŋɡjʊlə(r)]	adj. 非凡的；突出的；奇异的
bordered [ˈbɔː(r)də(r)d]	adj. 镶边的；有装饰边；织花边的
shroud [ʃraʊd]	n. 寿衣；裹尸布
gown [ɡaʊn]	n. 女长服；长外衣；外罩
embroider [ɪmˈbrɔɪdə(r)]	v. 刺绣；绣花
foliage [ˈfəʊliɪdʒ]	n.（植物的）叶；枝叶
sinuous [ˈsɪnjuəs]	adj. 弯曲有致的；蜿蜒的
gauze [ɡɔːz]	n. 薄纱，纱罗织物
brocade [brəˈkeɪd]	n. 织锦缎；（尤指用金银线织出凸纹的）厚织物
warp [wɔː(r)p]	n. 经纱

shed [ʃed]	n. 梭口
weft [weft]	n. 纬纱
treadle [ˈtred(ə)l]	n. (尤指旧时驱动机器的)踏板
auspicious [ɔːˈspɪʃəs]	adj. 吉利的；吉祥的
artefact [ˈɑː(r)tɪfækt]	n. 文物；人工制品，手工艺品(尤指有历史或文化价值的)
degum [diːˈɡʌm]	v. 脱胶；水洗
gum [ɡʌm]	n. 黏胶，胶质物(用以粘轻东西，如纸等)
relief [rɪˈliːf]	n. 浮雕
tapestry [ˈtæpɪstri]	n. 挂毯；织锦；壁毯；绣帷
motif [məʊˈtiːf]	n. 装饰图案；装饰图形
frontal [ˈfrʌnt(ə)l]	n. 额前装饰物(如发带、头帕)；(祭坛前面的)帷子
upholstery [ʌpˈhəʊlst(ə)ri]	n. 家具装饰品，坐垫和用来覆盖的织物
extant [ekˈstænt]	adj. 尚存的；现存的；未遭毁灭的
twill [twɪl]	n. 斜纹布
damask [ˈdæməsk]	n. 锦缎；花缎
revolution [revəˈluːʃ(ə)n]	n. 旋转，绕转
turnover [ˈtɜː(r)nəʊvə(r)]	n. 营业额，成交量

Critical reading and thinking

Task 1　Overview

Work in pairs and discuss the evolution of the ancient Chinese textile technology. Use as many lexical chunks as possible.

Task 2　Group discussion

Work in groups of 4-5 and have a discussion about the following questions.

1. Why does the jade sculpture of a human figure from Fuhao's tomb have a seemingly uncomfortable way of seating?
2. Some experts claim that museum displays of historical textiles can only partially represent

the clothing of our ancestors. Do you agree or disagree? Explain.

Task 3 Language building-up

Translate the following terms from English into Chinese or vice versa.

vegetable fibre	
bast fibre	
plain weave	
twill weave	
warp patterning	
weft patterning	
踞织机	
蹑机	
花机	
缠枝纹	

Task 4 Translation

I. Translate the following paragraph into Chinese.

Chinese textiles enjoy an excellent heritage in the textile sector and occupy a prominent position in the global textile market. Chinese textiles are world famous and extraordinary for their fine quality and profound symbolic meanings. Textiles in China often form an integral aspect of its heritage and symbolically reflect its tradition and culture. In China, the textile is often closely associated with prosperity and involved in the process of elaborate rituals. Parents' spontaneous love for their children is most visibly reflected through the excellent clothes they provide on festive occasions to their children. These clothes are made up of expensive materials and excellent craftsmanship.

II. Translate the following paragraphs into English.

语文是随着社会实践，特别是生产实践的发展而发展的。词汇往往反映在它初次出现之前久已普遍存在的社会现实。在汉族语文中，有大量的文字和词汇与纺织生产有关。在已经发现的甲骨文中，"糸"旁的字有一百多个。东汉人编的《说文解字》中所收"糸"旁的字有二百六十七个。还有"巾"字旁的七十五个，"衣"字旁的一百二十多个等都直接或间接与纺织有关系。至中华人民共和国成立前出版的《辞海》中所附的汉字读音表中所收的"糸"旁的字只有二百三十一个，可见我国纺织名词到汉代已相当完备，从侧面证明纺织技术在封建社会前期已经大体成熟了。

在现代汉语里,无论是各学科的术语,还是在常用的形容词、副词中,都有许多从纺织术语借用过来的字或词。就是从现代意义上已完全与纺织无关的一些抽象名词和成语中,也有不少是渊源于纺织的。例如"分析""成绩""综合""组织""纰漏""笼络人心""青出于蓝而胜于蓝"等,不胜枚举。这里,"分析""成绩"来源于纺麻;"综合""组织""纰漏"来源于织造;"络"来源于编结和缫丝;"青"和"蓝"来源于用植物染料染色。特别是"青出于蓝而胜于蓝"这句成语已流传了二千多年。

Task 5　Research

Surf the Internet for more information about the relation between computer programming and textile weaving.

(You can try https：//www. britannica. com/biography/Ada-Lovelace, https：//writings. stephenwolfram. com/2015/12/untangling-the-tale-of-ada-lovelace/.)

Section 2　Transmission

Lead-in

Warm-up questions：

A T-shirt may be readily available, but do you know how many disciplines are involved in the making? Complete the following checklist：

mathematics	dynamics	chemistry
mechanical engineering	electronics	materials science
manufacturing & systems engineering	industrial design	arts

Lexical chunk bank	
（英国）皇家特许状	royal charter
诺贝尔奖得主	Nobel laureate
介绍（纺织品）花色的小册子	pattern book
〈教〉非教派学院	non-sectarian college
红砖大学	redbrick university
中心辐射结构	hub and spoke structure
（美国）政府资助的低学费大学	land-grant college
（美国的）州议会	General Assembly
时装设计专业	fashion and apparel design
服务与纺织品管理专业	fashion and textile management

　　Textile technology evolved from handcrafts into family business, and after the Industrial Revolution, a **fully-fledged** industry with reliable spinning, weaving and finishing machinery in the 19th century. Clothing became basic necessities, and the textile industry expanded. This created a high demand of qualified professionals and executives in the textile centers like the cities of Leeds and Manchester. Weaving and spinning schools developed in

Europe in the middle of the 19th century and institutes of textile education sprung up around the world. Here we are going to introduce some of the key universities with textile threads in England, US, German, Russia, India, Japan, and China.

1. University of Leeds

The University of Leeds is a public research university in Leeds, West Yorkshire, England. It was established in 1874 as the Yorkshire College of Science. In 1884 it **merged** with the Leeds School of Medicine (established 1831) and was renamed Yorkshire College. It became part of the federal Victoria University in 1887, joining Owens College (which became the University of Manchester) and University College Liverpool (which became the University of Liverpool). In 1904 a **royal charter** was **granted** to the University of Leeds by King Edward VII.

The university has 36,250 students, the fifth largest university in the UK (out of 169). From 2006 to present, the university has consistently been ranked within the top 5 (alongside the University of Manchester, Manchester Metropolitan University, the University of Nottingham and the University of Edinburgh) in the United Kingdom for the number of applications received. Leeds had an income of £706.2 million in 2017/18, of which £137.1 million was from research grants and contracts. The university has financial endowments of £77.2 million (2017-2018), ranking outside the top ten British universities by financial endowment.

Notable **alumni** include current Leader of the Opposition Keir Starmer, former Secretary of State Jack Straw, former co-chairman of the Conservative Party Sayeeda Warsi, Piers Sellers (NASA astronaut) and six **Nobel laureates**.

The university's history is linked to the development of Leeds as an international center for the textile industry and clothing manufacture in the United Kingdom during the Victorian Era. The university's roots can be traced back to the formation of schools of medicine in English cities to serve the general public.

Before 1900, only six universities had been established in England and Wales: Oxford (founded c. 1096-1201), Cambridge (c. 1201), London (1836), Durham (1837), Victoria (1880), and Wales (1893).

The Victoria University was established in Manchester in 1880 as a federal university in the North of England, instead of the government elevating Owens College to a university and grant it a royal charter. Owens College was the sole college of Victoria University from 1880 to 1884; in 1887 Yorkshire College was the third to join the university.

The Yorkshire College of Science began by teaching experimental physics, mathematics, geology, mining, chemistry, and biology, and soon became well known as an international center for the study of engineering and textile technology (due to the manufacturing and textile trades being strong in the West Riding). When classics, modern literature and history went on offer a few years later, the Yorkshire College of Science became simply the Yorkshire College. In 1884, the Yorkshire College absorbed the Leeds School of Medicine and subsequently joined the federal Victoria University (established at Manchester in 1880) on 3 November 1887. Students in this period were awarded external degrees by the University of London.

The university library houses numerous **archives**, rare books and some objects in its Special Collections ranging from 2,500 BC to the 21st century.

The International Textile Collection (ITC) is made up of several distinct collections of world textiles, along with related objects, documents, and manuscripts. It dates from Ancient Egyptian to the present day, with the greater part covering the 19th and early 20th centuries.

Highlights of the collection include:
- ✓ European Fragments
- ✓ Egyptian Textiles
- ✓ Textiles of the Indian Subcontinent
- ✓ Indonesian Textiles
- ✓ Japanese Collection
- ✓ Kashmir Shawl Collection
- ✓ the Louisa Pesel Collection and Archive: collected (Mediterranean) and created textiles
- ✓ Qing Dynasty Textiles
- ✓ Sample and Pattern Books
- ✓ West African Textiles

The collection is particularly rich in embroideries from China and the Mediterranean Region, **resists** and weaves from Indonesia, India and West Africa, and European woven samples.

The British and French **pattern book** collection **charts** woven silks, velvets and wools from 1840s to 1970s, whilst the Louisa Pesel Collection is an outstanding archive of a remarkable educator in the art of embroidery. There are also small but notable collections of Egyptian children's garments, Tibetan *thangkas*, Japanese **stencils** and pattern books, and Kashmir **shawls**.

The collection will be of interest to researchers of fibre, technique, pattern, symbols, color, design history, religion, social history, manufacturing history, handicrafts, education, museum and cultural studies.

In the late 19th century professors in the Department of Textile Industries collected European fabrics in order to create pattern books as teaching resources for students of woven textile design. In 1892 a teaching museum was established, which was later described as the best of its type outside of London.

Local textiles firms, academics (including Head of Department Professor Aldred Barker and Vice Chancellor Sir Michael Sadler), students and associates continued to donate. With donations from China, India and Egypt, the collection soon took on an international theme.

By the 1960s this was a research collection within the Departmental Library. In 2004 an International Textiles Archive known as ULITA was established in St Wilfred's **Chapel**. This transferred to the Library in 2019. As an early established academic collection, it is now **exemplary** in containing many early and rare pieces.

2. University of Manchester

The University of Manchester traces its roots to the formation of the Mechanics' Institute (later UMIST) in 1824, and its heritage is linked to Manchester's pride in being the world's first industrial city. The English chemist John Dalton, together with Manchester businessmen and industrialists, established the Mechanics' Institute to ensure that workers could learn the basic principles of science.

John Owens, a textile merchant, left a **bequest** of £ 96,942 in 1846 (around £ 5.6 million in 2005 prices) to found a college to educate men on **non-sectarian lines**. His **trustees** established Owens College in 1851 in a house on the corner of Quay Street and Byrom Street which had been the home of the **philanthropist** Richard Cobden, and subsequently housed Manchester County Court. The **locomotive** designer, Charles Beyer became a **governor** of the college and was the largest single donor to the college extension fund, which raised the money to move to a new site and construct the main building now known as the John Owens building. He also campaigned and helped fund the engineering chair, the first applied science department in the north of England. He left the college the equivalent of £ 10 million in his will in 1876, at a time when it was in great financial difficulty. Beyer funded the total cost of construction of the Beyer building to house the biology and geology departments. His will also funded Engineering chairs and the Beyer Professor of Applied Mathematics.

The university has a rich German heritage. The Owens College Extension Movement based their plans after a tour of mainly German universities and polytechnics. Manchester mill owner, Thomas Ashton, chairman of the extension movement had studied at Heidelberg University. Sir Henry Roscoe also studied at Heidelberg under Robert Bunsen and they collaborated for many years on research projects. Roscoe promoted the German style of research led teaching that became the role model for the **redbrick universities** (英国19世纪初期开始用红砖建造的城市大学, 特指伦敦以外的). Charles Beyer studied at Dresden Academy Polytechnic. There were many Germans on the staff, including Carl Schorlemmer, Britain's first chair in organic chemistry, and Arthur Schuster, professor of Physics. There was even a German chapel on the campus.

In 1873 the college moved to new **premises** on Oxford Road, Chorlton-on-Medlock and from 1880 it was a constituent college of the federal Victoria University. The university was established and granted a Royal Charter in 1880 becoming England's first civic university; it was renamed the Victoria University of Manchester in 1903 and absorbed Owens College the following year. By 1905, the institutions were large and active forces. The Municipal College of Technology, forerunner of UMIST, was the Victoria University of Manchester's Faculty of Technology while continuing in parallel as a technical college offering advanced courses of study. Although UMIST achieved independent university status in 1955, the universities continued to work together. However, in the late-20th century, formal connections between the university and UMIST diminished and in 1994 most of the remaining institutional ties were severed as new legislation allowed UMIST to become an autonomous university with powers to award its own degrees. A decade later the development was reversed. The Victoria University of Manchester and the University of Manchester Institute of Science and Technology agreed to merge into a single institution in March 2003.

Before the merger, Victoria University of Manchester and UMIST counted 23 Nobel Prize winners amongst their former staff and students, with two further Nobel laureates being subsequently added. Manchester has traditionally been strong in the sciences; it is where the nuclear nature of the atom was discovered by Ernest Rutherford, and the world's first electronic stored-program computer was built at the university. Notable scientists associated with the university include physicists Ernest Rutherford, Osborne Reynolds, Niels Bohr, James Chadwick, Arthur Schuster, Hans Geiger, Ernest Marsden and Balfour Stewart. Contributions in other fields such as mathematics were made by Paul Erdös, Horace Lamb and Alan Turing and in philosophy by Samuel Alexander, Ludwig Wittgenstein and Alasdair MacIntyre. The author Anthony Burgess, Pritzker Prize and RIBA Stirling Prize-winning architect Norman Foster and composer Peter Maxwell Davies all attended, or worked at, Manchester.

The current University of Manchester was officially launched on 1 October 2004 when Queen Elizabeth **bestowed** its royal charter. The university was named the Sunday Times University of the Year in 2006 after winning the **inaugural** Times Higher Education Supplement University of the Year prize in 2005.

The founding president and vice-chancellor of the new university was Alan Gilbert, former Vice-Chancellor of the University of Melbourne, who retired at the end of the 2009-2010 academic year. His successor was Dame Nancy Rothwell, who had held a chair in physiology at the university since 1994. One of the university's aims stated in the Manchester 2015 Agenda is to be one of the top 25 universities in the world, following on from Alan Gilbert's aim to "establish it by 2015 among the 25 strongest research universities in the world on commonly accepted criteria of research excellence and performance". In 2011, four

Nobel laureates were on its staff: Andre Geim, Konstantin Novoselov, Sir John Sulston and Joseph E. Stiglitz.

The EPSRC (Engineering and Physical Sciences Research Council, 英国工程与自然科学研究理事会) announced in February 2012 the formation of the National Graphene Institute. The University of Manchester is the "single supplier invited to submit a proposal for funding the new £ 45m institute, £ 38m of which will be provided by the government"—(EPSRC & Technology Strategy Board). In 2013, an additional £ 23 million of funding from European Regional Development Fund was awarded to the institute taking investment to £ 61 million.

In August 2012, it was announced that the university's Faculty of Engineering and Physical Sciences had been chosen to be the "**hub**" location for a new BP (British Petroleum, 英国石油公司) International Centre for Advanced Materials, as part of a $ 100 million **initiative** to create industry-changing materials. The center will be aimed at advancing fundamental understanding and use of materials across a variety of oil and gas industrial applications and will be modelled on a **hub and spoke structure**, with the hub located at Manchester, and the spokes based at the University of Cambridge, Imperial College London, and the University of Illinois at Urbana-Champaign.

3. North Carolina State University

North Carolina State University (also referred to as NCSU, NC State, or just State) is a public **land-grant** research university in Raleigh, North Carolina. Founded in 1887 and part of the University of North Carolina system, it is the largest university in the Carolinas. The university forms one of the corners of the Research Triangle together with Duke University in Durham and the University of North Carolina at Chapel Hill. It is classified among "R1: Doctoral Universities—Very high research activity".

The North Carolina **General Assembly** established the North Carolina College of Agriculture and Mechanic Arts, now NC State, on March 7, 1887, originally as a land-grant college. Today, NC State has an enrollment of more than 35,000 students, making it among the largest in the country. NC State has historical strengths in engineering, statistics, agriculture, life sciences, textiles and design and offers bachelor's degrees in 106 fields of study. The graduate school offers master's degrees in 104 fields, doctoral degrees in 61 fields, and a Doctor of Veterinary Medicine.

NC State is also home to the only college dedicated to textiles in the country, the Wilson College of Textiles, which is a partner of the National Council of Textile Organizations (美国全国纺织品组织委员会) and is widely regarded as one of the best textiles **programs** in the world. In 2020 the textile engineering program was ranked 1st nationally by College Factual. In 2017, *Business of Fashion Magazine* ranked the college's **fashion and apparel**

design program 8th in the country and 30th in the world. In 2018, Fashion Schools ranked the college's **fashion and textile management** program 11th in the nation.

4. RWTH Aachen University

RWTH Aachen University or Rheinisch-Westfälische Technische Hochschule Aachen is a public research university located in Aachen, North Rhine-Westphalia, Germany. With more than 45,000 students enrolled in 144 study programs, it is the largest technical university in Germany.

In 2011, the university accounted for the highest amount of third-party funds of all German universities in both absolute and relative terms per faculty member. In 2007, RWTH Aachen was chosen by the DFG (Deutsche Forschungsgemeinschaft, 德国科学基金会) as one of nine German Universities of Excellence for its future concept RWTH 2020: Meeting Global Challenges and additionally won funding for one graduate school and three clusters of excellence.

RWTH Aachen is a founding member of IDEA League, a strategic alliance of five leading universities of technology in Europe. The university is also a member of TU9 (German Universities of Technology, 德国理工大学联盟), DFG and the Top Industrial Managers for Europe network.

RWTH Aachen University has educated several notable individuals, including some Nobel laureates in physics and chemistry. The scientists and alumni of the RWTH Aachen played a major role in chemistry, medicine, electrical, and mechanical engineering. For example, Nobel laureate Peter Debye received a degree in electrical engineering from RWTH Aachen and is known for the Debye model and Debye relaxation. Another example, Helmut Zahn and his team of the Institute for Textile Chemistry were the first who synthesized **insulin** in 1963 and they were nominated for Nobel Prize. Another example is B. J. Habibie, the third President of Indonesia that contributed in many aviation advancements. Franz Josef Och was the chief architect of Google Translate. Werner Tietz is one of the leading engineers of the Volkswagen Group and Vice President of SEAT.

5. Moscow State Textile University

A. N. Kosygin Moscow State Textile University was formed as Moscow State Textile Institute in 1919. It is one of the oldest institutes for higher studies in textiles in Russia.

In 1981, the institute was named in honor of Soviet Premier Alexei Kosygin, who died the previous year and whose profession was in the textile industry. The institute was upgraded to "Academy" in 1990. It was renamed to A. N. Kosygin Moscow State Textile Academy.

Nine years later, the Academy was approved as University and renamed as the A. N. Kosygin Moscow State Textile University in 1999.

The university has its own **complex**. It comprises 8 different campuses at the center of the city of Moscow, Russia. The teaching staff at the university is above 560, with 110 of them are Ph. D. and Professors.

The university has the following major departments:
➢ Technology and Production Management
➢ Chemical Technology and Ecology
➢ Weaving, Information Technology
➢ Automation and Energy
➢ Economics and Management
➢ Fashion Designing

The university offers specialization, masters and bachelors in 18 different categories. University has 41 departments, where it offers studies to almost 6,700 students. The university has 110 laboratories and 100 **auditoriums**. The university has its own sports hall, club and three **hostels**.

The University has served as a center of education for students from Russia and from all over the world. Many students from China, Pakistan, Morocco, Iran, Ghana, and India have completed their higher education at the university. The university has a one-room **mosque** which was built by Muslim students of the university in the hostel at Shablovskaya. The 7th floor of the university hostel is assigned to foreign students.

6. Indian Institute of Technology Delhi

Indian Institute of Technology Delhi (abbreviated as IIT Delhi) is a public technical and research university located in Hauz Khas, Delhi, India. It is one of the oldest Indian Institutes of Technology in India.

Established in 1961, IIT Delhi was formally inaugurated August 1961 by Prof. Humayun Kabir, Minister of Scientific Research & Cultural Affairs. First **admissions** were made in 1961. The current campus has an area of 320 acres (or 1.3 km^2) and is bounded by the Sri Aurobindo Marg on the east, the Jawaharlal Nehru University Complex on the west, the National Council of Educational Research and Training on the south, and the New Ring Road on the north, and flanked by Qutub Minar and the Hauz Khas monuments.

The institute was later **decreed** in Institutes of National Importance under the *Institutes of Technology Amendment Act*, 1963 and **accorded** the status of a full University with powers to decide its own academic policy, to conduct its own examinations, and to award its own degrees.

In 2018 IIT Delhi was also given the status of Institution of Eminence (IoE) by Government of India which granted almost-full autonomy. According to a government statement issued earlier, these IoEs will have greater autonomy in that they will be able to admit foreign students up to 30% of the admitted students and recruit foreign faculty up to 25% of the faculty strength with enhanced research funding.

The concept of IIT was first introduced by Sh. N. M. Sircar, then member of Education on **Viceroy**'s executive council. Following his recommendations, the first Indian Institute of Technology was established in the year 1950 in Kharagpur. In his report, Shri Sircar had suggested that such Institutes should also be started in different parts of the country. The Government, having accepted these recommendations of the Sircar Committee, decided to establish more Institutes of Technology with the assistance of friendly countries who were prepared to help. The first offer of help came from USSR (苏联) who agreed to collaborate in the establishment of an Institute through UNESCO at Bombay. This was followed by the Institutes of Technology at Madras, Kanpur and Delhi with collaborations with West Germany, United States and UK respectively. Indian Institute of Technology, Guwahati was established in 1994 and the University of Roorkee was converted into an IIT in 2001.

H. R. H. Prince Philip, Duke of Edinburgh, during his visit to India, laid the foundation stone of the college at Hauz Khas on 28 January 1959. The first admissions were made in 1961. The College of Engineering & Technology was registered as a society on 14 June 1960 under the *Societies Registration Act* No. XXI of 1860 (Registration No. S1663 of 1960-1961). The students were asked to report at the college on 16 August 1961, and the college was formally inaugurated on 17 August 1961 by Humayun Kabir, Minister of Scientific Research & Cultural Affairs. Initially, the college ran in the Kashmiri Gate campus of Delhi College of Engineering (now known as Delhi Technological University) before shifting to its permanent campus in Hauz Khas. The Department of Textile Technology of Delhi College of Engineering was shifted out **en bloc** to mark the beginning of the IIT Delhi at its new campus at Hauz Khas. The college was later accorded the status of a university and was renamed as Indian Institute of Technology Delhi. In 2018, IIT Delhi was one of the first six institutes to be awarded the Institute of Eminence status.

Department of Textile and Fibre Engineering

Founded in 1961, the department marked the establishment of IIT Delhi with a mission of conducting world-class research and providing quality education. It is the only department in the whole of IIT system and is one of the few worldwide with dedicated focus on teaching and research in textile and fibre engineering. The only department that **imparts** education and conducts research in all major areas of textiles vis. polymers & fibres, yarn, fabric, nonwovens and chemical processing.

With faculty having qualifications and research experience from the top international institutions, the department has strived to impart training and conduct research in the **emerging** areas of textile materials and engineering **at par** anywhere in the world.

Having contributed in creating new knowledge and developing new technologies, the department has positioned itself as one of the top institutions globally.

With faculty having qualifications and research experience from the top international institutions, the department has strived to impart training and conduct research in the emerging areas of textile materials and engineering at par anywhere in the world.

7. Shinshu University

Shinshu University (*Shinshū daigaku* 信州大学), abbreviated to *Shindai* (信大), is a Japanese national university in Nagano **Prefecture**(长野县), Japan. The only university of its established time, bearing the name "Shinshu" used before the 1871 establishment of prefectural name, is also firmly rooted in the many regions of Nagano Prefecture. It was the 18th ranked higher education institution in Japan.

The University has several campuses: Matsumoto(松元市), Nagano(长野县), Ueda(上田市) and Minami-Minowa(上伊那郡南箕轮村). Nagano Prefecture is located in the center of Japan, a region blessed with traditional culture that is kept alive at many places throughout the prefecture such as Zenkoji, Matsumoto Castle, Ueda Castle.

The school was founded in 1873 and it was established as a university in 1949, at which time the following institutions were subsumed into it: Nagano Normal School (established in 1873), Ueda Textile College (1910), Nagano Youths Normal School (1918), Matsumoto High School (1919), Nagano Technical College (1943), Matsumoto Medical College (1944), and Nagano Prefectural College of Agriculture and Forestry (1945).

History

Shinshu University traces its roots back to 1873 when it was a temporary normal school. The university was chartered by the Showa government in 1949 under a new Japanese education system reforming older system, merging seven institutions of higher education within Nagano Prefecture which includes Nagano Normal School (*Nagano shihan gakkō*, 长野师范学校) established in 1873, Ueda Textile College (*Ueda senni semmon gakkō*, 上田纤维专科学校) established in 1910, Nagano Youths Normal School (*Nagano seinen shihan gakkō*, 长野青年师范学校) established in 1918, Matsumoto Higher School (*Matsumoto kōtō gakkō*, 松元高等学校) established in 1919, Nagano Technical College (*Nagano kōgyō semmon gakkō*, 长野工业专科学校) established in 1943, Matsumoto Medical College (*Matsumoto ika daigaku*, 松元医科大学) established in 1944, and Nagano Prefectural College of Agriculture and Forestry (*Nagano kenritsu nōrin semmon gakkō*, 长野县立农林专科学校) established in 1945. In 2004 Shinshu University became a National University Corporation under the *National University Corporation Law*(《国立大学法人化法》).

Faculty of Textile Science and Technology

The Faculty of Textile Science and Technology is the only faculty in Japan with "Textile" in its name. It is recognized as an international research center in fibre engineering, and in 2010 celebrated the centenary of its establishment. It is ranked 35th in the sub-category "Textiles" in the international Science Citations Index. It is also ranked 50th in the "composites" sub-category. In 2012 they joined the Association of Universities for Textiles (AUTEX) based in Europe, to vitalize international exchange and collaboration.

8. Wuhan Textile University

Emerging at the historical juncture of rejuvenating China's national light industry, Wuhan Textile University (WTU) has gained a foothold in response to the demand of times, since the founding of its predecessor Wuhan Institute of Textile Engineering in 1958. In 1999, the institute was renamed as Wuhan University of Science and Engineering. In 2010, after successively merging the former Hubei Foreign Trade School and Hubei Academy of Finance and Economics, the present name of Wuhan Textile University was formally adopted.

By upholding the motto of "striving for truth and perfection" and advocating its campus spirit of "perseverance for cultivating qualified talents with unremitting efforts", WTU has been adhering to its path of characteristic development and open-up practice in education, vigorously supporting the textile industry for enhancing regional economy and social development. For the past 60-plus years, WTU has already established a wide range of disciplines in the fields of science, engineering, humanity, law, economics, management, arts, etc., with distinctive characteristics and advantages. Listed among the first batch of universities for "Basic Capacity Building Project for Universities in Western and Central China", "Education and Cultivation Scheme for Outstanding Engineers" initiated by Ministry of Education, PRC, "Domestic First-class Discipline Construction Initiative", as well as one of the "Top Ten Prestigious Fashion Institutes in China", WTU has been exerting on the construction of a high-caliber university featuring its graceful environment, outstanding spirits, adorable esteem among its faculties and students, with social prestige and distinctive characteristics in the field of textile industry.

Located in Wuhan city of Central China's Hubei Province, WTU has four campuses occupying an area of 130-plus hectares. Boasting 20 teaching departments and schools, it has more than 20,000 full-time students of various levels. Currently, it provides 66 programs for undergraduate students, including 4 featured programs of National-level, 3 programs of First-class Discipline Construction Initiative of National-level, 2 programs for accreditation of engineering education by Ministry of Education, PRC, 8 programs of "Education and Cultivation Scheme for Outstanding Engineers", as well as 7 provincial brand programs and

14 programs of the first-class disciplines of provincial level. With its 15 master's degree conferment programs of the first-class disciplines, 9 professional master's degree conferment programs and 7 discipline categories, WTU also boasts 8 provincial key disciplines of the first-level, plus 5 competitive and distinctive disciplines of provincial-level, among them including 2 discipline clusters (i. e., Modern Textile Technology, Fashion and Creative Culture) as key construction programs in the 13th Five-Year Plan for competitive and distinctive disciplines in provincial universities. The discipline textile science and engineering has been listed into "Domestic First-class Discipline Construction Initiative" in Hubei Province. The other three disciplines, such as material science, chemistry, and engineering, are listed into the front 1% of ESI's global ranking list.

By implementing the initiative of introducing high-caliber talents, WTU has become a magnet to converge the elites from all walks of life. The practice of adhering to both talents introduction and cultivation has enabled WTU to build a high-quality team of teaching faculty with lofty ideals and convictions, good morals, solid knowledge and in-depth learning, affection and benevolence. It is among the first batch of universities and colleges to be listed as the "Hubei Provincial Innovation and Entrepreneurship Base for Overseas High-caliber Talents". Currently, WTU boasts a faculty of 2,000-plus members, among them there are 1,300-plus full-time teachers, including those 700-plus teachers with senior professional titles and 700-plus teachers with doctoral degrees. In addition, we also boast a galaxy of top-notch elites, such as the seven academicians both from home and abroad who are working long-term for WTU (including one home-bred academician elected by Russian Academy of Natural Sciences), in addition to those 10 talents of national level, 200-plus talents of provincial level, 38 talents enjoying the special government allowance granted by the State Council, PRC, and 12 talents enjoying the specialized government allowance granted by Hubei Provincial People's Government. Besides that, we also boast a contingent of prominent talents entitled "sunshine scholars" with overseas educational backgrounds.

By implementing the scheme of cultivating qualified talents, WTU stands firm to the requirements of "four returns" proposed by Ministry of Education, with the philosophy of focusing on both moral cultivations and core elements of education. In the 3-year action plan of promoting the capacity of talent cultivation, we have established a sound paradigm of cultivating talents in moral education, intellectual education, physical education, aesthetic education and labor education as well. With the building of "output-oriented" talent cultivation system, we give full play to discipline construction featuring "novelty in both engineering and arts", so as to enhance the cultivation of practical and innovative talents for coordinated development in such aspects as "expertise, aptitude and morals". As the first batch of undergraduate university in enrollment work, WTU also boasts its teaching team and experimental zone for talent cultivation pattern reforms of national levels. Altogether, we have 2 first-class undergraduate courses of national levels and 6 national qualified open courses, in addition to 1 project for novelty research and practice in engineering approved by the Ministry of Education. In recent years, we have accomplished 2 second prizes of teaching results of national levels for universities and colleges, 11 first prizes of teaching results of provincial and ministerial levels. With strenuous efforts in implementing its education in innovation and entrepreneurship, WTU has been implementing the pattern of integrating both its innovation project and workshop. The establishment of Hubei Provincial Innovation and Entrepreneurship Base helps us to improve the students' capacity in innovation and practice as well as comprehensive qualities, with fruitful outcomes in discipline contests of national and provincial levels. By implementing the project of scientific and technological strengths, WTU attaches great importance to technological innovation for fostering key outcomes in scientific research. In this way, we aim at promoting transformation of these outcomes and incessantly scaling up our competitiveness in science and technology. We also boast a national key laboratory for new textile materials and advanced processing technology, a national-regional joint engineering laboratory, plus a key laboratory and an engineering research center supported by Ministry of Education, a national college heritage base for splendid traditional Chinese culture, as well as a batch of provincial key research bases. By undertaking nearly 200 projects sponsored by National Natural/ Social Science Foundation of China, including "973 Project", "863 Project" and some key supporting projects for scientific research of both national and provincial levels, WTU has gained distinctive advantages in the fields of textile, printing and dyeing, fashion and arts as well. Some of our achievements in scientific research have already been acknowledged as reaching advanced levels both internationally and domestically. In recent years, we has been awarded 1 first prize of National Science and Technology Progress Awards, 3 second prizes of National Science and Technology Progress Awards, in addition to the 2 second prizes of National Awards for Technological Invention. With joint efforts by WTU and the textile & fashion industry clustering regions to build "incubators" for production, education, and research, a batch of projects for collaborations between textile universities, enterprises, and regional governments have been put into practice in Hubei, Anhui, Jiangsu, Zhejiang, and Shandong. All these attempts have significantly helped us to scale up our capacity in social service.

With strenuous efforts in reinforcing international communication and cooperation, WTU attaches great importance to high-quality overseas education resources. All these are achieved by vigorously promoting education for international students and incessantly propelling its international-oriented approaches. WTU is the first member of AUTEX in Chinese mainland, the council member of TI in U. K. and the Fiber Society in U. S., initiator of the Belt and Road Alliance of Textile Higher Education, council institute of the Belt and Road Alliance of World Textile Universities, regional cooperative university of China-CEEC (17+1) Higher Education Institutions Alliance, supporting sector of Chinese Government Scholarship Entrusted Training Institution and Chinese Office of TI. WTU has established Research

Institute of International People-to-People Exchanges, the first of its kind by Ministry of Education, and Birmingham Institute of Fashion and Creative Art (BIFCA), the first Sino-overseas cooperation education institution of undergraduate level in Hubei Province. We have been conducting cooperative education programs of undergraduate level with University of Manchester, U. K., and Bunka Gakuen University, Japan. We have also been jointly conducting programs of cultivating doctoral students with universities in Australia, the United States, Russia, and Czech Republic. By signing international exchanges and cooperation agreements with nearly 200 universities in U. K., United States, France, Germany, Australia, Russia, Japan, and ROK, etc., we have long been undertaking the international seminars themed "Service Trade for Developing Countries" jointly sponsored by Ministry of Commerce, PRC, and United Nations Conference on Trade and Development. For consecutive years, WTU has made steady progresses in international communications by sponsoring and organizing different types of large-scale international academic conferences, with approvals on a package of programs themed "Lecturing Tour to Hubei Province by Prominent Scientists in the World".

Just as a citation of a poem in Tang Dynasty depicts, "The banks seem wide at the full tide; A sail with ease hangs in soft breeze." The same is also true for WTU's endeavors to keep pace with the development of the times and forge ahead in synergy with China's national industry to stand by its duty in social service for characteristic development, with its profound contributions to rejuvenate China's national industry as well as economic and social development in Hubei Province. In the 14th Five-Year Plan for National Economic and Social Development, WTU will keep a foothold on the phrase for new development, implement the ideal for new development, integrate itself into the paradigm of new development, and adhere to the systematic perspectives for overall planning. On the basis of cultivating people by moral education and student-centered ideals under the fully enhanced leadership of CPC, WTU will bring full play to further its overall reform, and propel its operation in accordance with laws and regulations for promoting the overall quality of talents. We will exert on cultivation of application and innovation-oriented talents by upgrading our capacity in scientific innovation and achievement transformation. With all these efforts, we will make more contributions to promote regional economy and social development as well as transformation and upgrading of the textile industry.

Words:

fully-fledged ['fʊliːfl'edʒd]	adj. 成熟的；羽毛丰满的；发育全的
merge [mɜː(r)dʒ]	v. 合并；融合；归并
grant [grɑːnt]	v. 允许；同意 n. (政府、机构的)拨款

alumnus [əˈlʌmnəs]	n. 毕业生；校友
archives [ˈɑːkaɪvz]	n. 档案；档案馆(室)
resist [rɪˈzɪst]	n. 防染布料
chart [tʃɑː(r)t]	v. 用图表示(说明)
stencil [ˈstens(ə)l]	n. (印文字或图案用的)模板；(油印)蜡纸
shawl [ʃɔːl]	n. 披肩；围巾
chapel [ˈtʃæp(ə)l]	n. (学校、监狱、私人宅院等基督教徒礼拜用的)小教堂
exemplary [ɪgˈzempləri]	adj. 典范的；可作榜样的；可作楷模的
bequest [bɪˈkwest]	n. 遗产；遗赠
trustee [trʌˈstiː]	n. 受托人；保管人
philanthropist [fɪˈlænθrəpɪst]	n. 慈善家；乐善好施的人
locomotive [ləʊkəˈməʊtɪv]	n. 机车；火车头
governor [ˈgʌvə(r)nə(r)]	n. (学校、学院、医院等机构的)董事，理事；负责人
polytechnic [pɒlɪˈteknɪk]	n. 综合性工艺学校(大学)；理工学院
premises [ˈpremɪsɪz]	n. 房产；(企业、机构的)营业场所
faculty [ˈfæk(ə)lti]	n. 全体教员；学院；系
bestow [bɪˈstəʊ]	v. 赐；授予；给予
inaugural [ɪˈnɔːgjʊrəl]	adj. 就职的；开幕的；成立的；创始的
hub [hʌb]	n. (某地或活动的)中心；枢纽
initiative [ɪˈnɪʃətɪv]	n. 倡议；新方案
program [ˈprəʊgræm]	n. 项目；计划；课程(表)；培养方案
insulin [ˈɪnsjʊlɪn]	n. 胰岛素
complex [kəmˈpleks]	n. (类型相似的)建筑群，综合体，大楼
auditorium [ɔːdɪˈtɔːriəm]	n. 大会堂；礼堂；〈美〉讲堂
hostel [ˈhɒst(ə)l]	n. 〈美〉(招待徒步旅行青年等的)招待所；〈英〉大学宿舍
mosque [mɒsk]	n. 清真寺

admission [əd'mɪʃ(ə)n]	n. 招生
decree [dɪ'kriː]	v. 裁定；判决；颁布（法令）
accord [ə'kɔːdɪd]	v. 给予，赠予，授予（权力、地位、某种待遇）
viceroy ['vaɪsrɔɪ]	n. （旧时受君主委派管治殖民地的）总督
en bloc [en blɔk]	adv. 整体；全部；一起；统统
impart [ɪm'pɑː(r)t]	v. 传授；告诉
emerging [ɪ'mɜː(r)dʒɪŋ]	adj. 新兴的；脱颖而出的；日益壮大的
at par (with)	adv. 与票面价值相等；媲美
prefecture ['priːfektʃə(r)]	n. 县；（法、意、日等国的）地方行政区域

Critical reading and thinking

Task 1 Overview

Work in pairs and discuss the development of textile education in Britain. Use as many lexical chunks as possible.

Task 2 Group discussion

Work in groups of 4-5 and have a discussion about the following questions.

1. What are the differences between the introductions of the English-speaking countries, such as those of the University of Leeds and the University of Manchester, and those of the Shinshu University and Wuhan Textile University?
2. How to improve the WTU introduction?

Task 3 Language building-up

Translate the following terms from English into Chinese or vice versa.

Vice Chancellor	
alumnus	
chapel	
governor	

auditorium	
at par (with)	
研究型大学	
研究经费	
诺贝尔奖得主	
中心辐射结构	

Task 4 Research

Surf the Internet for more information about the disciplines related to textile technology.

(You can try https://www.sciencedirect.com/topics/engineering/textile-science, http://textileconservation.academicblogs.co.uk/about-us/.)

Chapter 2 Textile History and Culture

Section 1 Textile History in Ancient China

Lead-in

Warm-up questions:

How much do you know about the historical development of Chinese textiles? Please identify the following pictures and match them with the corresponding descriptions.

(   ) (1) Silk trade between China and countries in Asia

(　　) (2) The earliest silk cocoon found in Yangshao Culture

(　　) (3) Leizu, the patroness of silkworms

(　　) (4) Huang Daopo, a pioneer and innovator of the cotton textile industry

Lexical chunk bank	
艺术纺织品	artistic textiles
经济支柱	economic staple
商业价值	commercial value
织造技术	weaving techniques
绸布	silk cloth
丝绸之路	the Silk Road
海上航线	sea routes
交换媒介	medium of exchange
跨洲贸易	transcontinental trade
棉纺织业	cotton textile industry
棉纺织机	cotton spinning frame
纺织品设计、制造工艺	craft of designing or creating textiles

1. Introduction

Textiles are fabrics or cloths and are one of the oldest forms of art practiced by many cultures. Perhaps no one, however, has appreciated the art form of textiles quite as long as the Chinese. Examples of truly **artistic textiles** going beyond simple rugs or shirts date back in China nearly to the Stone Age. Chinese textiles have a deep history in Asian culture but were actually amongst the most powerful forces in human history as well.

From the ancient times to the present day, people have been practicing the **craft of designing or creating textiles**. First emerging from a necessity to fill basic needs, methods of textile production have continually evolved. The textile making traditions spans global cultures as one of the earliest human technologies.

Silk production, characteristic of China's earliest civilization, has been an enduring feature of Chinese tradition and a distinctive aspect of China's interaction with other cultures. From China's **Neolithic Era**, hemp and ramie were cultivated and woven into textiles for clothing and other uses. **Wool textiles** played a minor role, associated with border peoples of the north and west. Cotton cultivation began at least by the eighth century. By the Ming

Dynasty (1368-1644), it had become an industry to rival silk production. Though silk retained its role as a luxury fabric and a symbol of Chinese culture, cotton cloth ultimately became a widespread material and an economic **staple**.

2. The legend of silk

Silk dominates Chinese textiles, and the Chinese have had a long time to master the use of this material. Few textile cultures are as defined by the material as the Chinese. That's because Chinese textiles are famously made of silk. Silk is produced from the **cocoons** of silkworms. The ancient Chinese had fully domesticated silkworms by the 4th millennium BC, but **archeologists** have also found silkworm cocoons in Stone Age sites as well. The oldest piece of **silk cloth** found was in China and dates to roughly 3630 BC.

Legend gives credit for developing silk to a Chinese empress, **Leizu**. A silkworm's cocoon fell into the teacup of the Empress Leizu. Wishing to **extract** it from her drink, the 14-year-old girl began to unroll the thread of the cocoon; seeing the long fibres that constituted the cocoon, the Empress decided to weave some of it, and so kept some of the cocoons to do so. Having observed the life of the silkworm on the recommendation of her husband, the Yellow Emperor, she began to instruct her entourage(随从) in the art of raising silkworms—**sericulture**. From this point, the girl became the goddess of silk in Chinese mythology. As the patroness of silkworms, she is still worshiped in China. While Leizu makes a proper and charming **patroness** of silk, it seems likely that silk did not come into much use until the Zhou Dynasty.

3. The history of silk textile

Silks were originally reserved for the Emperors of China for their own use and gifts to others, but spread gradually through Chinese culture and trade both geographically and

socially, and then to many regions of Asia. Because of its **texture** and **lustre**, silk rapidly became a popular luxury fabric in the many areas accessible to Chinese merchants. Silk was in great demand, and became a staple of pre-industrial international trade.

The earliest evidence of silk was found at the sites of Yangshao Culture(仰韶文化) in Xia County, Shanxi, where a silk cocoon was found cut in half by a sharp knife, dating back to between 4000 and 3000 BC. The species was identified as Bombyx mori(家蚕), the domesticated silkworm. Fragments of a primitive loom can also be seen from the sites of Hemudu Culture(河姆渡文化) in Yuyao, Zhejiang, dated to about 4000 BC. The earliest example of a woven silk fabric is from 3630 BC, and was used as wrapping for the body of a child. The fabric comes from a Yangshao site in Qingtaicun at Rongyang, Henan（河南荥阳青泰村）. Scraps of silk were found in a Liangzhu Culture Site at Qianshanyang in Huzhou, Zhejiang（浙江湖州千山阳良渚文化遗址）, dating back to 2700 BC. Other fragments have been recovered from royal tombs in the Shang Dynasty (c. 1600 BC-c. 1046 BC).

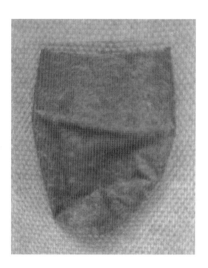

Findings from the Zhou culture confirm the early use of silk as a ground for painting, specifically in the second or third century BC used in funerary **rituals** and then buried with the deceased. A work known as the "Zhou Silk Manuscript（周氏帛书）", dated to circa 300 BC, documents the tradition that early Chinese texts were written on silk cloth and bamboo as well as cast in bronze or carved on stone. Examples of shop marks have been found on silks of this period, including a brocade with an impressed **seal**, suggesting a growing respect for distinctive workshop products and the **commercial value** of textiles.

Political disunity during the third to sixth centuries brought close interaction with Central Asia, leading to new styles and techniques relating to textile production. **Tang silks** reflect these closer contacts established during the previous centuries. The Tang maintained an open capital with foreigners among its merchants and varied ethnic and religious groups among its populace. A general shift in **weaving techniques** distinguishes Tang silk from that of the Han Dynasty. While Han patterns were warp-patterned, the weavings of Tang came to be weft patterned.

Elaborate techniques were developed for producing complex designs, both in the woven cloth itself and in **embellishments** worked onto the surface. Brocades, weaves with supplementary weft yarns creating complex patterns, were employed in many variations, including the complex lampas（彩花细锦缎）weave with its extra binding warps. Embroidery, a means of **embellishing** a woven fabric with **stitches** made using a threaded needle, flourished throughout the history of silk textiles in China.

Broadly speaking, the use of silk was regulated by a very precise code in China. For example, the Tang Dynasty and Song Dynasty imposed upon **bureaucrats** the use of particular colors according to their functions in society. Under the Ming Dynasty, silk began to be used in a series of accessories: handkerchiefs, wallets, belts, or even an embroidered piece of fabric displaying dozens of animals, real or mythical. These fashion **accessories** remained associated with a particular position: there was specific **headgear** for warriors, for judges, for nobles, and others for religious use. For more than a **millennium**, silk remained the principal diplomatic gift of the emperor of China to his neighbors or to his vassals(封臣).

For many centuries, silk was China's most valuable commodity. The spread of sericulture and the techniques of silk embroidery were so widespread throughout Ancient China that by the fifth century BC, more than 1/4 of the Chinese population were employed producing silk and creating silk embroideries. Throughout China many provinces became well known for their distinctive styles of embroidery.

4. The Silk Road

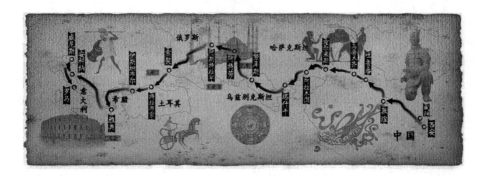

Numerous **archaeological** discoveries show that silk had become a luxury material appreciated in foreign countries well before the opening of the Silk Road by the Chinese. Though silk was exported to foreign countries in great amounts, the Emperors of China strove to keep knowledge of sericulture secret to maintain the Chinese monopoly. Consequently, other cultures developed their own accounts and legends as to the source of the fabric. In **classical antiquity**, most Romans, great admirers of the cloth, were convinced that the Chinese took the fabric from tree leaves.

The secret of silk was confined to China for around 300 years until the opening of the Silk Road in around 1000 BC. From the Han Dynasty, silk manufacture comprised a major state-controlled industry. For thousands of years, along land routes through Central Asia and **sea routes** along the coasts of East and Southeast Asia, silk was both a major commodity and at times a standard **medium of exchange**. In China's diplomacy, silk played a stabilizing role, bringing large areas of Inner Asia into the Chinese sphere of influence. At home, silk production was viewed as a moral **imperative** as well as a practical necessity. The Confucian **adage**, "men till, women weave", expresses the essential role of the women in a household in preparing silk yarn and cloth. Later, an increase in industrial specialization encouraged a shift of women's efforts from weaving to needlework, and yet the spirit of the phrase was preserved. As late as the seventeenth century, the state collected taxes in silk as well as in grain, underscoring the essential value of this human endeavor. Ultimately, the uses of silk expanded to include textiles made for appreciation as art as well as for clothing and furnishings.

Around 114 BC, the Han Dynasty, initiated the Silk Road Trade Route. Geographically, the Silk Road is an interconnected series of ancient trade routes between Chang'an (today's Xi'an) in China, with Asia Minor (小亚细亚) and the Mediterranean (地中海地区) extending over 8,000 km on land and sea. Trade on the Silk Road was a significant factor in the development of the great civilizations of China, Egypt, Mesopotamia (美索不达米亚), Persia (波斯), the Indian Subcontinent (次大陆), and Rome, and helped to lay the foundations for the modern world.

In the late Middle Ages, **transcontinental** trade over the land routes of the Silk Road declined as sea trade increased. The Silk Road was a significant factor in the development of the civilizations of China, India, Ancient Egypt, Persia, Arabia (阿拉伯半岛), and Ancient Rome. Though silk was certainly the major trade item from China, many other goods were traded, and various technologies, religions and philosophies, also traveled along the Silk Road. Some of the other goods traded included luxuries such as silk, **satin**, hemp and other fine fabrics, perfumes, spices, medicines, jewels, glassware, etc. China traded silk, teas, and **porcelain**, while India traded spices, ivory, textiles, precious stones, and pepper, and the Roman Empire exported gold, silver, fine glassware, wine, carpets, and jewels. Although the term "the Silk Road" implies a continuous journey, very few who traveled the route **traversed** it from end to end; for the most part, goods were transported by a series of agents on varying routes and were traded in the **bustling** markets of the oasis towns (绿洲).

5. The legend of cotton

It should be mentioned that silk was not the only material the ancient Chinese used. Cotton was introduced around 200 BC. Over the next several **millennia** it provided a stable source of cheaper textiles that were often treated with as much artistic **reverence** as silk.

Cotton textiles take cotton as material. In the southwest of China, the ethnic minority groups had cotton textiles early in the Eastern Han Dynasty. They called it "white folded cloth". In Fujian Province of the Han Dynasty, they grew cotton. In the northwest of China of the Three Kingdom Period they also had cotton textiles. The cotton textiles have been produced in the south of the Changjiang River since the Tang Dynasty, especially since the Yuan Dynasty.

By the late sixteenth century, cotton cultivation, which had been encouraged under the Yuan Dynasty and expanded further under the Ming Dynasty, became a major part of the Chinese economy. From at least Tang Dynasty times, cotton had been used to make clothing for the lower classes. In subsequent centuries, cotton cloth was associated with the virtues of humility.

Huang Daopo, a pioneer and innovator of the **cotton textile industry** in ancient China, helped transform China's cotton industry. She was born into a small village outside of Shanghai called Songjiang in about 1240 or 1245. She was sold at a very early age to another family as a child bride(童养媳). She was mistreated and so she ran away to Hainan Island.

While she was there, she learned a whole new style of weaving. At the age of 50, she returned to Songjiang, bringing with her this new weaving style and a new type of cotton seed that was easier to grow. The seeds and innovations she brought back with her helped Songjiang become the cotton manufacturing capital of the world. Europe wouldn't see similar technologies until the 1700s.

The most important innovation of Huang Daopo was her **three-spool, pedal-driven cotton-spinning machine**. Before, local people used the **single-spool hand wheel** to spin **cotton roving** into workable yarn or thread. Huang Daopo's **multi-spool machine** was the most advanced **cotton spinning frame** in the world at that time. It was invented more than 400 years before **the Spinning Jenny** was created by James Hargreaves in England in 1764.

Domestically, cotton found wide use in **undergarments** and in **linings** for silk (such as in silk garments for ceremonial use), and it also came to be dyed in bright colors and **calendered** to an attractively polished surface. In the areas populated by the ethnic minority groups the cotton used to serve as the material which was made into cotton textiles of various colors called brocades. The term "brocade" here refers to textile fabric and textile variety made with different weaving skills. There are silk-knit goods and cotton goods. Some textiles are woven with the blending of silk and cotton. The most distinctive artistic developments in Chinese cotton textiles are those that survive primarily in rural traditions and folk art relating

to the ethnic minority groups, including the Miao of Guizhou and Yunnan Provinces. The primary techniques, resist dyeing (using **stencils** to apply a paste that would retain white undyed areas), and **batik** (using wax to reserve undyed areas), had been known in China, along with **block printing**, **tie-dyeing**, and clamp-resist. The characteristic blue from **indigo**, typical of dyed cotton, also reflects an ancient tradition, recorded in detail in Ming Dynasty texts.

Words:

staple [ˈsteɪpl]	n. 主要产品，支柱产品
cocoon [kəˈkuːn]	n. 蚕茧
archeologist [ˌɑːkɪˈɒlədʒɪst]	n. 考古学家
archaeological [ˌɑːkiəˈlɒdʒɪkl]	adj. 考古学的
extract [ˈekstrækt]	v. 提取，取出
sericulture [ˈserɪˌkʌltʃə]	n. 养蚕
patroness [ˌpeɪtrəˈnes]	n. （女）守护神
texture [ˈtekstʃə(r)]	n. 质地；手感
lustre [ˈlʌstə(r)]	n. 光泽
ritual [ˈrɪtʃuəl]	n. 仪式，典礼
seal [siːl]	n. 印章，标志
embellish [ɪmˈbelɪʃ]	v. 装饰；修饰
embellishment [ɪmˈbelɪʃmənt]	n. 装饰品
embroidery [ɪmˈbrɔɪdəri]	n. 刺绣
stitch [stɪtʃ]	n. 针法；针脚

bureaucrat [ˈbjʊərəkræt]	n. 官僚
accessory [əkˈsesəri]	n. 配件，配饰
millennium [mɪˈleniəm]	n. 千年期
millennia [mɪˈleniə]	n. 千年期（millennium 复数）
headgear [ˈhedɡɪə(r)]	n. 帽子
imperative [ɪmˈperətɪv]	n. 重要紧急的事；必要的事
adage [ˈædɪdʒ]	n. 格言
transcontinental [ˌtrænzˌkɒntɪˈnentl]	adj. 横贯大陆的
satin [ˈsætɪn]	n. 缎子
porcelain [ˈpɔːsəlɪn]	n. 瓷器
traverse [ˈtrævɜːs]	v. 穿过
bustling [ˈbʌslɪŋ]	adj. 熙熙攘攘的，忙乱的
reverence [ˈrevərəns]	n. 崇敬
undergarment [ˈʌndəɡɑːrmənt]	n. 内衣
lining [ˈlaɪnɪŋ]	n. 内衬；衬里；内胆
calender [ˈkælɪndə]	v. 用砑光机压光
stencil [ˈstensl]	n. 蜡纸；模板
batik [bəˈtiːk]	n. 蜡染
indigo [ˈɪndɪɡəʊ]	adj. 靛蓝；靛青

Phrases:

Neolithic Era	新石器时代
wool textiles	羊毛纺织品
Tang silks	唐代丝绸
classical antiquity	古典时代
three-spool, pedal-driven cotton-spinning machine	脚踏式三轴棉纺机
single-spool hand wheel	单轴手轮
cotton roving	面粗纱
multi-spool machine	多轴纺纱机

the Spinning Jenny	珍妮纺纱机
block printing	手工木板印花
tie-dyeing	扎染

Proper names:

Leizu	嫘祖

Critical reading and thinking

Task 1 Overview

Work in pairs and retell the legend of silk, the Silk Road, and the legend of cotton according to the following pictures. Use as many lexical chunks as possible.

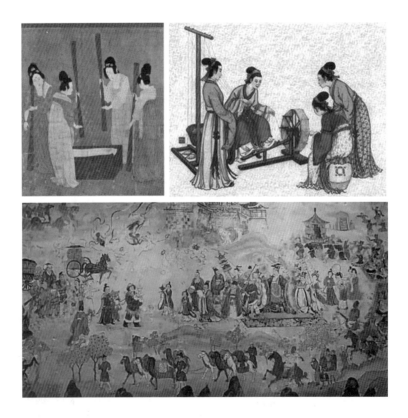

Task 2 Group discussion

Work in groups of 4-5 and have a discussion about the following questions.

1. Please compare the similarities and major differences between the legend of silk and the legend of cotton.
2. What are the contributions to the development of the society and economy during the history of textiles in China?

Task 3 Language building-up

I. Translate the following terms from English into Chinese or vice versa.

silk cloth	
artistic textiles	
economic staple	
bustling markets	
spinning frame	
棉花种植	
织造技术	
海上航线	
多轴纺纱机	
棉纺织业	

II. Complete the following sentences with the words or phrases given in the previous exercise.

1. The early Chinese texts were written on _____ and bamboo as well as cast in bronze or carved on stone.
2. Though silk retained its role as a luxury fabric and a symbol of Chinese culture, cotton cloth ultimately became a widespread material and an _____.
3. A general shift in _____ distinguishes Tang silk from that of the Han Dynasty. While Han patterns were warp-patterned, the weavings of Tang came to be weft patterned.
4. For the most part of "the Silk Road", goods were transported by a series of agents on varying routes and were traded in the _____ of the oasis towns.
5. Huang Daopo, a pioneer and innovator of the _____ in ancient China, who helped transform China's cotton industry.

Task 4 Translation

I. Translate the following paragraph into Chinese.

Silk is a natural fibre, some forms of which can be woven into textiles. The protein fibre of silk is composed mainly of fibroin(蚕丝蛋白) and produced by certain insect larvae(幼虫) to form cocoons. The best-known type of silk is obtained from the cocoons of the larvae of the mulberry silkworm Bombyx mori reared in captivity (sericulture). The shimmering appearance of silk is due to the triangular prism-like structure of the silk fibre, which allows silk cloth to refract incoming light at different angles, thus producing different colors. The first evidence of the silk trade is the finding of silk in the hair of an Egyptian mummy of the 21st Dynasty, 1070 BC. The silk trade reached as far as the Indian Subcontinent, the Middle East, Europe, and North Africa. The trade was so extensive that the major set of trade routes between Europe and Asia came to be known as the Silk Road. In the ancient era, silk from China was the most lucrative and sought-after luxury item traded across the Eurasian Continent, and many civilizations, such as the ancient Persians, benefited economically from trade.

II. Translate the following paragraph into English.

丝绸原产中国，最早的例证历史可以追溯至公元前 3500 年。丝绸最早被中国的皇帝自己使用，或赏赐他人，之后随着中华文化的传播而被作为地理乃至社会意义上的商品，出口至亚洲各国。由于其良好的质地和鲜亮的光泽，丝绸作为一种极受欢迎而又奢侈的物品走入中国商人的眼中。很快，丝绸变成畅销品，丝绸业也成为前工业国际贸易的支柱产业。

Task 5 Research

Surf the Internet for more information about the history of textile in China.

(You can try https：//fashion-history. lovetoknow. com/fabrics-fibres/chinese-textiles, https：//study. com/academy/lesson/ancient-chinese-textiles. html.)

Section 2　Textile and Culture

Lead-in

Warm-up questions:

1. How much do you know about the relation between textile and culture during the Chinese history? Please try to translate the following quotes and discuss their relations with the textiles.
（1）东门之池，可以沤麻。东门之池，可以沤苎。东门之池，可以沤菅。
　　　　　　　　　　　　　　　　　　　　——《诗经·陈风·东门之池》
（2）迢迢牵牛星，皎皎河汉女。纤纤擢素手，札札弄机杼。
　　　　　　　　　　　　　　　　　　　　——《古诗十九首》
（3）十三能织素，十四学裁衣，十五弹箜篌，十六诵诗书。十七为君妇，心中常苦悲。君既为府吏，守节情不移。贱妾留空房，相见常日稀。鸡鸣入机织，夜夜不得息。三日断五匹，大人故嫌迟。
　　　　　　　　　　　　　　　　　　　　——《孔雀东南飞》
（4）长安一片月，万户捣衣声。秋风吹不尽，总是玉关情。
　　　　　　　　　　　　　　　　　　　　——《子夜吴歌·秋歌》

2. Do you know the representative brocade and embroidery skills in China? Please identify the following pictures and match them with the corresponding descriptions.

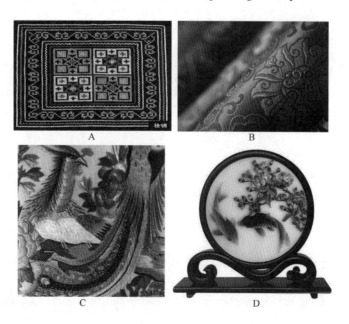

(　　) (1) Shu Brocade, historical silk-knit brocade with a general term for the silk-knit the flower brocades.
(　　) (2) The Zhuang Brocade of Guangxi, a traditional handcraft with more than 2,000 years of history.
(　　) (3) Su Embroidery, embroidery with fish on one side and kitty on the other side is a representative of this style.
(　　) (4) Yue Embroidery, a general name for embroidery products of the regions in Guangdong Province. The most famous piece is "Hundreds of Birds Worshiping Phoenix".

Lexical chunk bank	
文化转型	cultural transformation
诗歌流派	poetry genre
文学气质	literary temperament
唐绣艺术	Tang embroidery art
刺绣流派	schools of embroidery
民族文化	ethnic cultures
服饰制度	apparel system
刺绣工艺	craft of embroidery
原始织机	primitive looms
云锦	Cloud Brocade
湘绣	Xiang Embroidery
文化遗产	cultural heritage

1. Textile and *Han Fu*

The Qin and Han Dynasties were a period of social and **cultural transformation** in China. After the **Warring States Period** of the beacon fire smoke, writers absorbed the Warring States Period of prose and *Chu Ci* writing techniques, developed a new **poetry genre** in the history of Chinese literature— *Han Fu*(汉赋). In the poems written by writers in the Qin and Han Dynasties, there are a lot of textile production techniques and textile trade, as well as a wealth of social and folk culture information.

The cultivation and textile production of mulberry and flax in the Qin and Han Dynasties

were closely related to the high attention of the ruling class. In the Pre-Qin Period, the rulers of the past dynasties regarded heavy agriculture as a way to enrich country and strengthen army. The Han Dynasty was the earliest dynasty in the history of Chinese agriculture to raise mulberry production to the same status as agricultural production. It can be seen from *Han Fu* that textile production and foreign trade in the Han Dynasty were important pillars of economy.

The early Western Han Dynasty was a development period of *Han Fu*, because of influence of *Chu Ci* leaving the style of poetry, the writing technique of *Han Fu* at this time is exaggerated, length is longer, representing **literary temperament**. The middle period in the Han Dynasty was the peak period of *Han Fu*. At this time, *Han Fu* was of great length, mostly describing the scenery of Kyoto(京都) in the Han Dynasty and eulogizing the prestige in the Han Dynasty. The late Han Dynasty was the transition period of *Han Fu*. At this time, *Han Fu* was short in length, lyrical in scenery, fresh in artistic conception and strong in rhythm. It reflected return of the style in the late Han Dynasty to the Pre-Qin Period.

Han Fu contains rich contents of ancient Chinese textile science and technology. These poems are the direct historical data of development of textile industry and new direction of research on the development of textile technology and society in the Han Dynasty. There are many kinds of textiles in the Han Dynasty, most of which are reflected in *Han Fu*. *The First Aid* (《急就章》) written by Shi You(史游), a calligrapher, was a primer for children's literacy in the Han Dynasty, more than 50 kinds of silk are recorded in it. The textile content recorded in it has clearly described silk fabric, according to the fabric structure, pattern, color, processing technology into four categories. Shi You's article not only praised the superb skill level of the silk weaving industry, but also reflected the prosperity of the silk weaving industry. It shows that the prosperity of textile culture is high, and textile production has been integrated into every aspect of people's daily life.

The Han Dynasty was the peak period for the development of textile technology in ancient China. Around the origin of textile materials, the Han Dynasty formed several important textile production centers such as Shandong, Sichuan and Henan. Flourishing textile trade promoted the rapid improvement of the craft level of silk and hemp products and wool textiles. Developed silk weaving industry promoted the wide range of textile tools such as **reeling**, spinning wheels and foot-tilting looms(脚踏斜织机). According to the records of *The Notes on West Capital*(《西京杂记》), in the early years of the Western Han Dynasty, the wife of Chen Baoguang, a giant deer man, improved the traditional **flat heald jacquard** process into **bunched heald jacquard**, which greatly improved the efficiency of brocade. According to the description of *Fu on Weaving Women*(《妇赋》) written by Wang Yi in the Eastern Han Dynasty, the jacquard machine in the Han Dynasty had a combination of **fuselage** and installation system, which basically had the main components of traditional jacquard machine in China.

The textile production in the Han Dynasty was based on individual families and small-scale private ownership of producers, which was a self-sufficient **small-scale peasant economy**. Small producers can not only satisfy the daily life of the family, but also trade the **surplus goods** in the family production. The rulers of the Han Dynasty fixed the family life pattern of men ploughing and women weaving by law and policy. Small-scale peasant economy became the **template** of family life in ancient China and was inherited by subsequent dynasties. It was the implementation of the policy of combining agriculture and mulberry cultivation by the rulers that pushed textile economic development to the peak in the Han Dynasty. With Zhang Qian's exploration of the Western regions, silk from **the central plains** was continuously exported to Central Asia and introduced to Europe, creating the world-renowned textile economy of the Silk Road.

Mulberry and hemp were widely planted in the Han Dynasty, and there were many kinds of hemp and silk fabrics. The silks were colorful and of **exquisite** quality. With the opening of the Silk Road, the fine silk fabrics were sold to other countries in the Western regions, which made the textile economy prosperous in the Han Dynasty. The hemp fabric is soft and fine in texture and cheap in quality. In the Han Dynasty, the clothes of the common people were mostly made of **linen**. The extravagant price of silk goods determined that only the upper ruling class could use it. The civilians have no right to wear silk, even if economic conditions permit. The common people had to meet the requirements of the dress and **apparel system** in order to be allowed to wear clothes made of silk in their old age.

Although there is no special chapter about costumes in *Han Fu*, there are many depictions related to costumes because of its complicated content. According to *Han Fu*, the emperor's clothing in the Han Dynasty had twelve patterns of embroidery, and the **regalia** was decorated with jade, seal and sword.

The opening of the Silk Road promoted the economic and trade prosperity and development of the regions along the Silk Road. It has deepened the textile trade and cultural exchanges between China and the West. It promoted friendly politics, and people-to-people exchanges between the Han Dynasty and the countries in the Western regions. It can be seen from the names of the Western regions and various exotic foreign objects appearing constantly in *Han Fu* that in the Han Dynasty, the countries in the Western regions had close economic and cultural exchanges with the central plains. The strange customs and goods of the Western regions recorded in *Han Fu* fully illustrate the prosperity of the textile economy and the frequent foreign economic and trade exchanges in the Han Dynasty. The Silk Road promoted the high prosperity in the Han Dynasty's textile economy. The textiles of Han Dynasty were popular in many countries in the Western regions. It shows the superb textile technology in ancient China and the sericulture civilization with a long history.

After the Silk Road was opened, the Western Han Dynasty's economic development increased rapidly. The flourishing commerce via the Silk Road increased the wealth of Han society and improved the quality of life of its people. At that time, Chang'an, the capital of the Han Dynasty, was the most prosperous city in the world, with a highly developed social economy and prosperous villages and towns. Goods from five prosperous textile centers are gathered here and sold in the Western regions and throughout the country. In *Shu Du fu* (《蜀都赋》), Yang Xiong described the busy commercial scene in Chengdu. The four sides of the merchants gathered in the land of Shu, the noisy market, everywhere was the voice of the peddlers one after another.

While the commerce and trade in the Western regions were flourishing, the commercial business of the maritime Silk Road was no less than that of the land route. The production of silkworm mulberry in the Qin and Han Dynasties was the beginning of China's ancient textile economy. According to the records of *Han Fu*, the highly developed textile business in the Han Dynasty led to the prosperity of social economy. Economic and trade exchanges via the Silk Road not only enhanced the cultural exchanges between China and the West, but also opened the integration of Chinese and the people of the West.

2. Textile and art

Textiles have been a fundamental part of human life since the beginning of civilization. The silk tradition in China dates back to around 3639 BC, and it involves elaborately patterned brocades, complex **gauze** weaves, and intricately embroidered textiles. With artistry and technical accomplishment that still amaze modern viewers, the demand for

Chinese silk textiles was very high in many distant lands of the time.

Fabrics made of silk consist of many types: brocade, satin, silk fabric, etc. This variety is due to different weaving skills and silk fabrics. Some are lined, some are **unbleached**, some are heavy, and some are thin. **Silk-knit goods** are one of great Chinese contributions to the world culture. The weaving skills emerged in the primitive society. They can demonstrate the culture tradition of one nation. Though they historically served as clothing material, its relation to the common people had never been severed. Many excellent weaving skills and patterns were first established by the common people and passed to all walks of life.

2.1 Brocade

Brocade is a type of silk-knit goods whose patterns are highlighted by the colorful horizontal silk. First the horizontal threads are installed on the common weaving machine. Under the horizontal threads there are colorful picture drafts. The vertical threads with various colors are woven in segment by the small **shuttles** according to the patterns. The horizontal thread of each color is interwoven with the vertical thread with every other color. This way of weaving is called "interweaving horizontal and vertical threads". China's "Four Famous Brocades" were once famous all over the world for their exquisite patterns and complicated weaving techniques.

2.1.1 Shu Brocade

Shu Brocade is one of historical silk-knit brocade and a general term for the silk-knit the flower brocades which were manufactured in Chengdu, Sichuan Province, from the Han Dynasty to the Three Kingdoms Period. The earliest written record of Shu Brocade is in the Spring and Autumn Period, which has lasted for more than 2,000 years. Shu Brocade is a kind of multi-color brocade with the characteristics of Han ethnic group and local style. It flourished until the Tang, Song and Yuan Dynasties. Of the Sichuan brocades in the Tang Dynasty, the bundle flower lining brocade and the red lion and phoenix lining brocade were the most outstanding. Sichuan brocade is based on horizontally colored line. As a carrier of cultural exchange and trade, the world-famous "Silk Road" between North and South has witnessed history and unique cultural value.

2.1.2 Song Brocade

Song Brocade, Suzhou Brocade, is the traditional silk-knit brocade in Suzhou, Jiangsu Province. It appeared in the Southern Song Dynasty with exquisite texture and unique skills, which was used not only for clothing and costumes, but also for mounting calligraphy and painting scrolls. There were more than 40 kinds of brocade. Especially the application of **mounted calligraphy and painting**, so that these beautiful and luxurious brocade and painting treasures can be preserved together. It was then lost at the end of the Ming Dynasty, and recovered at the beginning of the Qing Dynasty. It consists of big brocade and small brocade. Among them the big brocade is also called heavy brocade, which is mainly used for mounting picture and decoration, while small brocade is used for making box and decorating small articles. They are patterned geometrically and neatly decorated with bundles of flowers and flowers on twigs.

Song Brocade, with its gorgeous color, exquisite design and soft texture, is the "crown of rich brocade" in China. In 2012, China succeeded in restoring the production of Song Brocade on modern looms. On the basis of retaining traditional Song Brocade skills, China continued to incorporate fashion elements, and made innovations and breakthroughs in product quality, types and styles. Song Brocade products not only have been presented as national gifts for many times to politicians of various countries, but also showcased in APEC, G20 and other international conferences, Milan Fashion Week and other international fashion arenas.

2.1.3 Cloud Brocade

Cloud Brocade, or Nanjing Brocade, was developed during the Yuan Dynasty although its origin could date back to the Southern Dynasty. Among all ancient fabrics, silk cloth known as brocade, jin, represents the top arts and crafts of the industry. Furthermore, Nanjing Brocade has absorbed all the best silk fabric weaving crafts and skills of past dynasties and ranks first in quality among the Shu Brocade in southwestern Sichuan Province, Song Brocade in Jiangsu Province, and Zhuang Brocade in southwestern Guangxi Province. It is named after its color as gorgeous as colorful cloud, for it is made of high quality silk and

woven with exquisite skill. The silk industry consists of two trades: the pattern brocade trade and the unpatterned brocade trade since the end of the Qing Dynasty. Not until then the name "cloud brocade" came into use.

2.1.4 Zhuang Brocade

The Zhuang Brocade of Guangxi is a traditional handcraft with more than 2,000 years of history, also becoming the most vivid interpretation of the aesthetics of a nation. The vivid brocade pattern, rigorous structure and colorful colors highlight the warm and cheerful style of the Zhuang people as the crystallization of the wisdom of Zhuang people and a synonym for preciousness.

Produced by local Zhuang people, Zhuang Brocade is a splendid handicraft which originated in the period of the Tang and Song Dynasty. Woven with cotton threads and colorful silk threads, the Zhuang Brocade was the special gift offered by the local government to the royal family in the Ming Dynasty. Zhuang Brocade is favored by people for its beautiful patterns, which show a unique Zhuang style and favor, and its durable quality. Also, other characteristics like wide-ranging themes, well-knit structure, vivid designs, exquisite patterns and rich colors reflect the moral characters of bravery and industry, wisdom and sensitiveness, as well as honesty and frankness of Zhuang people. Typical patterns on the Zhuang Brocade include unique shapes such as letters, water, squares, clouds, flowers and so on. Zhuang Brocade has many uses, such as blankets, quilt facings, **aprons**, bags, girdles, scarves, cloth borders and wall hangings.

2.2 Embroidery

The birth and gradual improvement of silk making led to the birth of embroidery, which, as a regional handcraft, adopted distinct regional and ethnic character. In the Spring and Autumn Period, the art of embroidery was already quite mature. Themes of embroidery in this period include exaggerated figures of dragon, phoenix and tiger, **interspersed** with flowers, foliage or geometrical shapes. Entering the Qin and Han Dynasties, embroidery reached a new height. The workshop set up in Linzi, capital of the Qi Kingdom, dedicated exclusively to the making of official court uniforms. Thousands of weavers were hired. Not only the royal family had the whole house dress in brocade, but all the rich wore what was called the "five colored brocade" and decorated their furniture with silks and embroideries. One great achievement of the **Tang embroidery art** is the invention of a new stitch—the satin stitch(绷针), which is popular until the present day. This new stitch brought more freedom to the artist, and brought about a new era of embroidery. The Song Dynasty marks the peak of Chinese embroidery, both in terms of quality and quantity. In the Ming and later Qing Dynasties, embroidery reached its peak in popularity. Local **schools of embroidery** appeared, the most famous being Suzhou, Guangdong, Sichuan, Hunan, Beijing and Shandong schools. In addition to their local flavors, these schools all borrowed from other **ethnic cultures**. Embroidery is an important treasure of the Chinese traditional arts and crafts and is an important part of Chinese nation. It represents the wisdom of Chinese people. Fortunately, the art and **craft of embroidery** have been preserved as China's great **cultural heritage**.

2.2.1 Su Embroidery

Su Embroidery, the general name for embroidery products in areas around Suzhou, Jiangsu Province. Su is the short name for Suzhou. A typical southern water town, Suzhou and everything from it reflects tranquility, refinement, and elegance. So does Su Embroidery.

Embroidery with fish on one side and kitty on the other side is a representative of this style.

Favored with the advantaged climate, Suzhou with its surrounding areas is suitable for raising silk and planting mulberry trees. As early as the Song Dynasty, Su Embroidery was already well known for its elegance and vividness. In the Ming Dynasty, influenced by the Wu School of painting, it began to rival painting and calligraphy in its artistry.

In history, Su Embroidery dominated the royal wardrobe and walls. Even today, it occupies a large share of the market in China as well as in the world.

2.2.2 Shu Embroidery

Originated from Shu, the short name for Sichuan, Shu Embroidery, influenced by its geographic environment and local customs, is characterized by a refined and brisk style. The earliest record of Shu Embroidery was during the Western Han Dynasty. At that time, embroidered product was a luxury enjoyed only by the royal family and was strictly controlled by the government. During the Han Dynasty and the Three Kingdoms Period, Shu Embroidery and Shu Brocade were exchanged for horses and used to settle debts.

In the Qing Dynasty, Shu Embroidery entered the market, and an industry was formed. Workshops and governmental bureaus were fully devoted to it, promoting the development of the industry. It became more elegant and covered a wider range. From the paintings by masters, to patterns by designers, to landscape, flowers and birds, dragon and phoenix, tiles and ancient coins, it seemed all could be the topic of embroidery. Folk stories like the "Eight Immortals Crossing the Sea", "Kylin(麒麟) Presenting a Son" and other auspicious patterns such as magpie on plum and mandarin ducks(鸳鸯) playing on the water were also favorite topics. Patterns with strong local features were very popular among foreigners at that time. These local features included lotus and carp, bamboo forest and pandas. Some bought embroidered skirts and used them as curtains.

2.2.3 Xiang Embroidery

As art from Hunan, it was a witness of the ancient Xiang (Hunan) and Chu (Hubei) culture. It was a gift to the royal family during the Spring and Autumn Period. The most

persuasive evidence is the articles unearthed in Mawangdui Han Tomb.

Developing over two thousand years, Xiang Embroidery became a special branch of the local art. It gained popularity day by day. Besides the common topics seen in other styles, it absorbed elements from calligraphy, painting and inscription.

Its uniqueness is that it is patterned after a painting draft but is not limited by it. Perhaps because of this technique, a flower seems to send off fragrance, a bird seems to sing, a tiger seems to run, and a person seems to breathe.

2.2.4 Yue Embroidery

Yue embroidery is a general name for embroidery products of the regions in Guangdong Province. According to historical records, it encompasses Embroidery of Guangzhou and Chaozhou, and has the same origin as **Li Brocade**. People generally agree that it started from the Tang Dynasty since Lu Meiniang(卢媚娘), who embroidered seven chapters of Buddhist sutra(佛经), was from Guangdong. Portraits, flowers and birds are the most popular themes as the subtropical climate favors the area with abundant these plants that are rarely seen in central China. In addition, it uses rich colors for strong contrast and a magnificent and bustling effect.

Since Cantonese take to fortunes in an almost superstitious attitude, attaching a lucky implication to everything, red and green, and auspicious patterns are widely used. The most famous piece is "Hundreds of Birds Worshiping Phoenix". Fish, lobsters, bergamots and lychee are also common patterns.

2.2.5 Gu Embroidery

Gu Embroidery distinguishes itself from other local styles by the fact it originated from Gu Mingshi's family during the Ming Dynasty in Shanghai, instead of from a certain place. It is also known as Lu Xiang Yuan Embroidery. Lu Xiang Yuan, Dew Fragrance Garden in Chinese, was where the Gu Family lived. From the start, it was different from other styles as it specialized in painting and calligraphy. The inventor was a concubine of Gu Mingshi's first son, Gu Huihai. Later, Han Ximeng(韩希孟), the wife of the second grandson of Gu Mingshi developed the skill and was reputed as "Saint Needle"(针圣). Some of her masterpieces are kept in the Forbidden City. Today it has become a special local product in Shanghai.

2.2.6 Bian Embroidery

Bian Embroidery was regarded as a National Treasure during the Northern Song Dynasty. Bian refers to the capital of the Northern Song Dynasty, Bianliang, today's Kaifeng. It was mainly used by the royal family, so it was also known as Court or Official Embroidery. The style was exquisite, precise and elegant to match the demeanor of the royal family. However, with the collapse of the dynasty, this technique collapsed, too.

2.2.7 Han Embroidery

Han Embroidery originated from Chu (Hubei Province) and flew to Wuhan from Jingzhou and Shashi. Tinted by the Chu Culture, it is characterized by a rich and gaudy color with bold patterns and exaggerated techniques. It came to its heyday in the middle and later Qing Dynasty and obtained golden medals in international expos and competitions. Embroidery Street was formed in Daxing Road, Hankou, with nearly 40 workshops engaged in it. Bombing by the American planes of a Japanese magazine nearby destroyed the street as weavers fled.

2.2.8 Embroidery by Ethnic Minority Groups

Among ethnic minority groups, Bai, Bouyei and Miao people are also adept at embroidery. Their embroidered products use sharp contrast of color and primitive design to express a mysterious flavor while embroidered Thangka by Tibetans shows their passion in religion.

2.3 The future of textile and art

The history of textile is also interwoven with the history of international trade, as items such as the Tyrian purple dye(提尔紫染料), Chinese silk and other luxury fabrics were important trade goods among countries. The history of textile is very rich and complex and traditions such as weaving, tapestry making, fabric dying, embroidery or quilting are ancient. Today, artists are using these methods to create new and exciting works.

Ever since the 1980s, textile arts have been developing new forms and languages involving many creatives along the way. Influenced by postmodernist ideas, textile and fibre work has become more and more conceptual. Various creatives are now experimenting with techniques, materials and concepts, completely pushing the limits of the medium. These reborn practices such as embroidery art, weaving, quilting, crochet and many others, have placed a new focus on the work that confronted social and political issues such as gender feminism, domesticity, women's work, and identity politics. Yet not all fibre artists are feminists or even concerned with the political and social connotations of fabric arts. They simply employ textiles and threads as a painting and sculpting material. Today, contemporary pieces explore a variety of textile and fibre practices and techniques that provide **a myriad of** possibilities. All of the creatives and many more continue to build the important legacy of different fabric arts reinventing the medium in various ways.

Words:

reel [riːl]	v. 卷，绕上卷轴
heald [hiːld]	n. 综；综框；综线
fuselage [ˈfjuːzəlɑːʒ]	n. 机身
template [ˈempleɪt]	n. 模板
exquisite [ɪkˈskwɪzɪt]	adj. 精致的
linen [ˈlɪnɪn]	n. 亚麻布
regalia [rɪˈɡeɪliə]	n. 正式场合象征地位的服饰
gauze [ɡɔːz]	n. 薄纱
unbleached [ʌnˈbliːtʃt]	adj. 原色的，未经漂白的
shuttle [ˈʃʌtl]	n. 梭；梭子
scroll [skrəʊl]	n. 长卷纸，卷轴
apron [ˈeɪprən]	n. 围裙
interspersed [ɪntəˈspɜːst]	adj. 点缀的

Chapter 2 Textile History and Culture

Phrases:

Warring States Period	战国时代
flat heald jacquard	水平织机，平综提花机
bunched heald jacquard	束综提花机
small-scale peasant economy	小农经济
surplus goods	剩余商品
the central plains	中原地区
silk-knit goods	针织品
mount calligraphy and painting	书画装裱
Shu Brocade	蜀锦
Song Brocade	宋锦
Zhuang Brocade	壮锦
Su Embroidery	苏绣
Shu Embroidery	蜀绣
Yue Embroidery	粤绣
Li Brocade	黎锦
Gu Embroidery	顾绣
Bian Embroidery	汴绣
Han Embroidery	汉绣
a myriad of	大量的

Critical reading and thinking

Task 1 Overview

Work in pairs and retell the relation between textile and culture, such as textile and *Han Fu*, and textile and art according to the following pictures. Use as many lexical chunks as possible.

Section 2 Textile and Culture

Task 2　Group discussion

Work in groups of 4-5 and have a discussion about the following questions.

1. Do you know some other literature works, such as poems and proses, concerning the textiles skills or textile industry in the history of China? Please share your collections with your group members.
2. What are the differences between the brocades and embroideries? Please summarize the major features of the famous brocades and embroideries.

Task 3　Language building-up

Ⅰ. **Translate the following terms from English into Chinese or vice versa.**

apparel system	
poetry genre	

literary temperament	
silk-knit goods	
bunched heald jacquard	
精美的花纹	
小农经济	
剩余商品	
文化遗产	
书画装裱	

II. Complete the following sentences with the words or phrases given in the previous exercise.

1. After the Warring States Period of the beacon fire smoke, writers absorbed the Warring States Period of prose and *Chu Ci* writing techniques, developed a new _____ in the history of Chinese literature— *Han Fu*.
2. The civilians have no right to wear silk, even if economic conditions permit in the Han Dynasty. The common people had to meet the requirements of the dress and _____ in order to be allowed to wear clothes made of silk in their old age.
3. The textile production in the Han Dynasty was based on individual families and small-scale private ownership of producers, which was a self-sufficient _____.
4. Brocade is a type of _____ whose patterns are highlighted by the colorful horizontal silk. China's "Four Famous Brocades" were once famous all over the world for their exquisite patterns and complicated weaving techniques.
5. Embroidery is an important treasure of the Chinese traditional arts and crafts and is an important part of Chinese nation. It represents the wisdom of Chinese people. Fortunately, the art and craft of embroidery have been preserved as China's great _____.

Task 4 Translation

I. Translate the following paragraph into Chinese.

China has a wealth of classical literature, both poetry and prose, dating from the Eastern Zhou Dynasty (770-256 BC). The proponents of the Hundred Schools of Thought in the Spring and Autumn and Warring States Periods made important contributions to Chinese prose style. Among the earliest and most influential poetic anthologies was the *Chuci*. The songs in this collection are more lyrical and romantic and represent a different tradition from the earlier *Shijing*. During the Han Dynasty (202 BC-220 AD), this form evolved into the *fu*, a literary form combining elements of poetry and prose. The form developed during the Han Dynasty

(202 BC-220 AD) from its origins in the long poem *Lisao* by Qu Yuan (c. 340 BC-c. 278 BC). The *fu* was particularly suitable for description and exposition, in contrast to the more subjective, lyrical *sao*. Its prosody was freer than that of the *sao*, the rhyme pattern being less restrictive.

Ⅱ. **Translate the following paragraph into English.**

中国是丝绸的故乡，因而有很多与丝绸相关的艺术，刺绣就是其中一种。刺绣是中国民间传统手工艺之一，至少有两三千年的历史。从事刺绣的多为女子，因此刺绣又被称为"女红"。刺绣在中国受到人们的广泛喜爱。刺绣可以用来装饰衣物，如在衣服、被子、枕套等物品上绣上美丽的图案，也可制作成特别的饰品。中国的四大名绣：苏州的苏绣、广东的粤绣、湖南的湘绣以及四川的蜀绣。各种绣法不仅风格有差异，主题也各有不同。其中，苏州的苏绣最负盛名。

Task 5 Research

Surf the Internet for more information about the textile and culture in China.
(You can try https：//www. britannica. com/art/fu-Chinese-literature, https：//study. com/academy/lesson/ancient-chinese-textiles. html.)

Chapter 3　Textile and Life

Section 1　About the Clothing

Lead-in

Warm-up questions：

1. Why do people wear clothes?
2. What roles do you think clothes play in our life from the ancient times to the modern times?
3. What do you know about the history of clothing?

Lexical chunk bank	
全套连衣裙	full dress
宽大的连衣裙	voluminous dresses
粗呢长连衣裙	thick woolen long dresses
轻薄棉衣	light cotton clothing
金属胸针	metal brooches
裹腿裤	leg wrappers
软花边鞋	soft laced shoes
合成纤维	synthetic fibres
缝制皮革	sewn leather
天然纤维	natural fibres
毛皮服装	fur garments

机织纤维	woven fibres
高腰连衣裙	high-waisted dresses
亚麻布	linen cloth
装饰别针	ornamented pins
宽松半身裙	voluminous skirt
成衣时尚	ready-to-wear fashion
简单束腰外衣	simple tunic

1. What is clothing?

Clothing, also known as clothes, **apparel** and **attire**, is the collective term used to describe the different types of materials worn on the body. Clothing is typically made of fabrics or textiles but over time has included **garments** made from animal skin or other thin sheets of materials put together. The wearing of clothing is mostly restricted to human beings and is a feature of all human societies. The amount and type of clothing worn depends on gender, body type, social and geographic considerations. Clothes assume significance only when they are on the body. Clothes are more than just products of a textile factory or exhibits in a museum; they are **artifacts**, used by people in all activities of daily life. Their true significance only becomes apparent when we consider how they are related and adapted to the body.

2. Why wear clothing?

People wear clothes for many reasons, primarily for protection and decoration.

Generally, clothing serves many purposes: It can help protect us from various types of weather, and can improve safety during hazardous activities such as hiking and cooking. It protects the wearer from rough surfaces, **rash-causing plants**, insect bites, **splinters**, **thorns** and **prickles** by providing a barrier between the skin and the environment. Clothes can insulate against cold or heat. They can also provide a **hygienic** barrier, keeping infectious and toxic materials away from the body. Clothing also provides protection from harmful **UV radiation**.

The most obvious function of clothing is to improve the comfort of the wearer, by protecting the wearer from the elements. In hot climates, clothing provides protection from sunburn or wind damage, while in cold climates its **thermal insulation properties** are generally more important. Shelter usually reduces the functional need for clothing. For example, coats, hats, gloves, and other superficial layers are normally removed when entering a warm home, particularly if one is residing or sleeping there. Similarly, clothing has seasonal and regional aspects, so that thinner materials and fewer layers of clothing are generally worn in warmer seasons and regions than in colder ones.

Clothing performs a range of social and cultural functions, such as individual, occupational and sexual differentiation, and social status. In many societies, norms about clothing reflect standards of modesty, religion, gender, and social status. Clothing may also function as a form of adornment and an expression of personal taste or style.

Some clothing protects from specific environmental hazards, such as insects, noxious chemicals, weather, weapons, and contact with **abrasive substances**. Conversely, clothing may protect the environment from the clothing wearer, as with doctors wearing medical scrubs.

3. The history of clothing

It is not certain when people first started wearing clothes, however, anthropologists estimate that it was somewhere between 100, 000 and 500, 000 years ago. The first clothes were made from natural elements: animal skin, fur, grass, leaves, bone, and shells. Garments were often **draped** or tied; however, simple needles made out of animal bone provide evidence of sewn leather and fur garments from at least 30, 000 years ago.

When settled **neolithic cultures** discovered the advantages of woven fibres over animal hides, the making of cloth, drawing on basketry techniques, emerged as one of humankind's fundamental technologies. Hand and hand with the history of clothing goes the history of textiles. Humans had to invent weaving, spinning, tools, and the other techniques needed to be able to make the fabrics used for clothing.

Ancient civilizations like Greece and Rome favored wide, unsewn lengths of fabric from which they constructed their clothing (fabric was expensive and they didn't want to cut it). Ancient Greek clothing was made of lengths of rectangular wool or linen cloth which was secured at the shoulders with ornamented pins and belted with a **sash**. Women wore loose robe called **peplos**, men cloak called **chlamys** while both men and women wore **chiton**—a type of tunic which was short to the knees for men and longer for women.

The **toga** of ancient Rome which was worn by free Roman men citizens was also an unsewn length of wool cloth. Under the toga they wore a simple tunic which was made from two simple rectangles joined at the shoulders and sides. Roman women wore the draped **stola** or a tunic that had length to the ground.

During the Iron Age that lasted from 1200 BC to 500 AD women of northwestern Europe wore wool dresses, tunics and skirts which were held in place with leather belts and metal **brooches** or pins. Men wore **breeches** with leg wrappers for protections and long trousers. They also wore caps and **shawls** made from animal skin and soft laced shoes made from leather.

During Medieval times the **Byzantines** made and exported very richly patterned cloth. Expensive variant was woven and embroidered while cheaper, intended for lower classes was resist-dyed and printed. They wore tunics, or long chitons over which they wore **dalmatica**, which is a heavier and shorter type of tunics or long cloaks.

At the same time look of European clothing depended on whether people who wore it identified with the old Romanized population or the new invaders such are Franks, Anglo-Saxons or Visigoths. Men of the invading peoples wore short tunics with belts and visible trousers, hose or leggings. In 12th and 13th century Europe clothing remained simple. In 13th century dyeing and working of wool improves and Crusaders bring with them craft of silk. Fashion begins in Europe in 14th century.

In Renaissance Europe wool remained the most popular fabric for all classes but the linen and hemp were also used. More complex clothes were made and urban middle class joins the fashion that was set by higher class and royalties. Early Modern Europe from 16th century sees even more complex fashion with ruffs, **passementerie** and needle lace. Enlightenment introduces two types of clothing: "full dress" worn at Court and for formal occasions, and "undress" which are everyday, daytime clothes. Full dress almost disappeared by the end of the 18th century.

Industrial Revolution brings machines that spin, weave and sew and with that produce fabric that is of better quality, faster made and has lower price. Production moves from small cottage production to fabrics with assembly lines.

20th century invents synthetic fibres that is cheaper than natural and which is mixed with many natural fibres.

4. Clothing and fashion

Fashion history is the history of people. People have loved clothing fashion for thousands of years. From the early days of Egypt to the day clothes have become the expression of who we are. Throughout history, drawings, documents and other archaeological finds have also revealed fashion worn by people in various ancient civilizations.

The Greeks were wearing clothes in different styles from their eastern neighbors. Both women and men were usually wearing thick woolen long dresses.

The ancient Egyptians were typically dressed in light cotton clothing. Women were wearing long, ready-made clothes. With the Romans becoming dominant over time, the Romans began to be known as the most important example of style and fashion at the same time.

With the beginning of migration to Europe towards the north, the styles changed. Women used more material to keep them warm and men used armor or plain clothes.

Since there was no way or material to produce new clothes for those who migrated to America, clothes were ordered and sent to the ocean by ships. Men were wearing suits with leggings, and women were always wearing long clothes and wearing hats at their heads. Forms continued to transfer from **voluminous** dresses into soft, frequent, high-waisted dresses.

Godey's *Lady's Book* was the most fashionable voice of the 1800's. In the following period they praised French fashion and became very popular in the United States. After 1700 stylists returned to the more voluminous skirt. The **synthetic paints** allowed the clothes to be painted in various colors. This opened the door to new, brighter styles. Up to this point, all the clothes were made specially, but **Abba Gould Woolson** started to create **ready-to-wear fashion**, where the clothes would be produced in bulk and sold to the masses.

At the beginning of the century, women's skirts became thinner and shorter. Men's trousers slowly went from knee to ankle length. Between World Wars, women's fashion often turned into shorter skirts just below the knee. In men, Levi's jeans have become increasingly popular.

The 1960s and 1970s are great transitional periods in fashion. Women wearing trousers has seen more and more accepted. The clothes became shorter and looser. 80's brought great fashion trends and different hairstyles for both men and women.

As time went by from the 20th century to the 21st, fashion continued to change every year, and today still changing.

Words:

apparel [əˈpærəl]	n. 服装；衣服
attire [əˈtaɪə(r)]	n. 服装；盛装
garment [ˈɡɑːmənt]	n. 衣服，服装
artifact [ˈɑːtɪfækt]	n. 人工制品，手工艺品
splinter [ˈsplɪntə(r)]	n. 碎片；微小的东西
thorn [θɔːn]	n. 刺；荆棘
prickle [ˈprɪkl]	n. 刺；刺痛；植物的皮刺
hygienic [haɪˈdʒiːnɪk]	adj. 卫生的，保健的；卫生学的
scrubs [skrʌbz]	n. 外科手术服
drape [dreɪp]	vt. 用布帘覆盖；(使)呈褶裥状
sash [sæʃ]	n. 腰带；肩带
peplos [ˈpeplɒs]	n. (古希腊的)女式长外衣；女式大披肩

chlamys [ˈklæmɪs]	n. 古希腊男子所着的一种短斗篷或外套
chiton [ˈkaɪtn]	n. 希顿古装(古希腊人贴身穿的宽大长袍)
toga [ˈtəʊgə]	n. (古罗马的)宽外袍
stola [ˈstəʊlə]	n. 女士及踝长外衣
brooch [brəʊtʃ]	n. (女用的)胸针,领针
breech [briːtʃ]	n. 臀部;后膛
shawl [ʃɔːl]	n. 围巾,披肩
dalmatica [dælˈmætɪkə]	n. 教士法衣
passementerie [ˈpæsm(ə)ntrɪ]	n. 珠缀;衣服的金银饰带
voluminous [vəˈluːmɪnəs]	adj. 大的;宽松的

Phrases:

rash-causing plants	引起皮疹的植物
UV radiation	紫外线;紫外辐射
thermal insulation properties	隔热性能
abrasive substance	研磨材料
neolithic culture	新石器文化
synthetic paints	合成油漆
ready-to-wear fashion	成衣时装

Proper names:

Byzantines	拜占庭人
Godey's *Lady's Book*	戈迪的《女书》(19世纪美国的一本杂志)
Abba Gould Woolson	乌尔森(美国的女性存在主义教母)

Critical reading and thinking

Task 1 Overview

1. Work in pairs and retell the reasons why people wear clothes. Use as many lexical chunks as possible.
2. Please retell the history of clothing.

Task 2 Group discussion

Work in groups of 4-5 and have a discussion about the following questions.

1. What characteristics of clothing were exhibited during the different historical times?
2. Please identify the differences of clothing fashions in different periods of the world history.

Task 3 Language building-up

Ⅰ. **Translate the following terms from English into Chinese or vice versa.**

synthetic fibres	
leg wrappers	
high-waisted dresses	
neolithic culture	
thermal insulation properties	
金属胸针	
机织纤维	
天然纤维	
引起皮疹的植物	
宽松半身裙	

Ⅱ. **Complete the following sentences with the words or phrases given in the previous exercise.**

1. _____ are fibres made by humans through chemical synthesis, as opposed to natural fibres that are directly derived from living organisms.
2. Discovered in the 1950s, the Qujialing Site is the most representative site of _____ in

the middle reaches of Yangtse River. It was selected as one of the first 100 large site-conservation projects by the National Cultural Heritage Administration in 2005.

3. Adorable maxi skirts, _____ will all be on your shopping list as you get ready for the season.

4. Little babies should be kept away from _____ which will cause their skins to feel uncomfortable.

5. During the Iron Age that lasted from 1200 BC to 500 AD women of northwestern Europe wore wool dresses, tunics and skirts which were held in place with leather belts and _____ or pins.

Task 4　Translation

Ⅰ. Translate the following paragraphs into Chinese.

The history of Traditional Chinese Clothing here will introduce the brief history of the development of Chinese costume. China has many ethnic groups with a long history while Han people dominate most periods in history. For thousands of years, generations of clothing designers have devoted themselves to building the Kingdom of Clothes, making the garments that cover the human body into an important component of Chinese culture. The progress of nation can be seen through its changes in clothing styles.

Clothing manufacture in China dates back to prehistoric times, at least 7,000 years ago. Archaeological findings of 18,000 year-old artifacts such as bone sewing needles and stone beads and shells with holes bored in them attest to the existence of ornamentation and of sewing extremely early in Chinese civilization.

Ⅱ. Translate the following paragraphs into English.

在中国古代社会，服装是地位和职业的象征。富人和穷人的着装在各个方面都不同。富人穿着丝绸制成的衣服，而穷人穿着大麻制成的衣服。皇帝穿黄色的衣服，而隋朝的穷人只允许穿蓝色或黑色。衣服的颜色是情感的象征，红色系象征幸福，白色系象征早晨。让我们来探索中国传统服饰的风格和历史。

在中国古代文化中，个人在社会中的地位在很大程度上决定了所穿衣服的种类。限制范围从裙子的长度、袖子的宽度以及装饰量而定。在这段时间里，时尚通常非常简单和中性。中国的时尚是由朝代发展而来的，尽管如此，其服装非常有限。

Task 5　Research

Surf the Internet for more information about the clothing history.

(You can try https：//en. wikipedia. org/wiki/History_of_clothing_and_textiles, http：//www. historyofclothing. com/clothing-history/.)

Section 2　Textiles and Clothing

Lead-in

Warm-up questions:

1. When you buy clothes, what elements of clothing do you usually consider?
2. Can you easily identify what materials the clothes are made from? What fabrics do you know?
3. Would the style of clothing influence your choice of clothes?

Lexical chunk bank	
绉纱	crepe yarns
轻质织物	lightweight fabric
半合成织物	semi-synthetic fabric
蓝白方格布	blue and white gingham
透气织物	breathable fabric
软帘雪纺半身裙	soft drapery chiffon skirt
Z 捻绉纱	z-twist crepe yarns
纬线	weft thread
单经线	single warp thread
棉混纺织物	cotton blend fabric
机织染色纱	dyed yarn woven
棉混纺	cotton blends
平纹针织物	jersey knit fabric
轻薄单罗纹针织	slight single rib knit
亚麻织物	linen fabric
印染织物	printed and dyed fabrics

花式编织条纹	fancy woven stripe
丝织锦缎	woven silk brocade
醋酸酯平纹针织衫	acetate jersey
真丝织物	silk fabric
羊毛混纺	wool blend
羊绒纤维	cashmere fibres
平纹机织物	plain-woven fabric
暖色调	warm colors
流行色预测	fashion color predictions
最强视觉冲击力	strongest visual impact
时尚色彩	fashion colors
冷色	cool colors
深色收缩	dark colors contract
色彩象征	color symbolism
现有调色板	existing color palette
藏青蓝	navy blue
舒缓的色彩	soothing color

Textiles, to archaeologists anyway, can mean woven cloth, bags, nets, basketry, string-making, cord impressions in pots, sandals, or other objects created out of organic fibres.

When we want to gain a deep insight into textiles, four items should have taken into the consideration: texture, pattern, color, and fabric.

1. Texture

Texture is the surface quality of an object. In fashion design, the texture is the surface interest of a fabric, created by the weave and by light reflection. Our eyes can appreciate the play of light on smooth or rough surfaces and the hands to feel the surface. Combinations of textures such as leather with chunky knit (glossy with scratchy) create excitement in a garment. All fabric textures, from the sheerest chiffon to bulkiest **fleece** to the sturdiest canvas, depend on the variations of four factors: fibre content, **yarn** structure, fabric structure and **finishes**. All these factors affect the visual, tactile, and performance quality of a texture. Nowadays, texture is one of the key elements in the current fashion trend and plays an important role in the whole visual effect of a garment and its appearance.

Texture influences the **drape** of a garment. Chiffon clings and flows, making it a good choice for soft, feminine styles while canvas has the firmness and bulk suitable for more casual garments.

Texture affects the color of a fabric by causing the surface to either reflect or absorb light. Rough textures absorb light, causing the colors to appear flat. Smooth textures reflect light, causing colors to appear brighter. A color that appears extremely bright in a shiny **vinyl** or **satin** will become subdued in rough wool or suede.

Texture is regarded as one of the significant components in the fashion history. Wonderful effects can be achieved when texture is introduced to a garment of a single color. This can be implemented by decoration, pin-tucking, **smocking**, gathering and embroidery—anything that disturbs the surface.

In fact, a majority of designers select fabrics before making their design sketches; they prefer to be inspired by the texture rather than to find the perfect fit for a design sketch. Furthermore, texture can also create the illusionary effect of narrowness and fullness in the wearer, just as the way lines, patterns and colors achieve other illusionary effects. Texture can affect the appearance of a **silhouette**, giving it a bulky or a slender look, depending on the roughness or smoothness of the materials. The contrast in texture between the stiff top and the **soft drapery chiffon skirt** by Martin creates a strong impact for this fashionable texture outfit.

2. Patterns

Fabric patterns are coordinated by lines, dots, shapes, spaces and colors. They are created in endless varieties—geometric, floral, stripes, checkers, dots, abstract **motifs** and many others. Consequently, patterns always provide interest and visual impact. Printed and dyed fabrics convey new fashion trends easily and comparatively quickly. They help balance collections or ranges and add variety. Their pattern arrangement can be subtle or intense, large or tiny, bright or dark, even or uneven and spaced or clustered. The ideas come from worldwide influences including nature motifs, folkloric, historic, architecture, wallpapers, cartoons and people's hobbies. Designers may shop worldwide for fabric patterns or garments as a source of new idea. The most common and traditional fabric patterns are divided into four categories and are listed as follows:

2.1 Geometric patterns

Geometric patterns refer to textural patterns, stripes, checks and plaids that are woven, printed or knitted such as **gingham**, left-hand **twill**, striped **chambray**, window pane

linen or **pin stripe**.

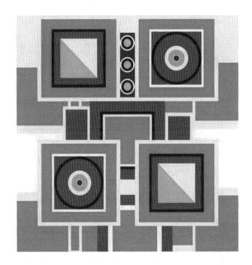

2.2 Conventional patterns

Conventional patterns refer to naturalistic motifs that are stylized. Conventional patterns combine the rhythm of stripes with the soft natural charm of floral in pattern. For example, fancy woven stripe on **dimity**, woven silk brocade or block print on linen.

2.3 Naturalistic patterns

Principally floral, although other motifs may be used, ranging from leopard spots to candy canes such as **paisley** on **acetate jersey**, naturalistic floral on cotton and stylized floral on silk **crepe**.

2.4 Dots and spots

For example, novelty dots on cotton, coin dots on flat crepe, and geometric motifs on jersey.

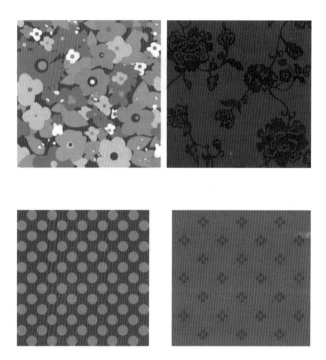

3. Colors

Colors have always been a key element in women's clothing and a fundamental consideration in the design process as they are the very first element to catch shoppers' gaze and they are the last thing to **clinch** a purchase. It is often the first element that is noticed with a design and it influences how that garment or collection is perceived. Thus, colors are often the staring point of the design process.

Colors can be considered to have the strongest visual impact among other design related elements on an object. For example, warm colors and pure colors appear nearer to the viewer whereas cool colors appear to recede. Light colors expand; dark colors contract. Yellow is perceived as the largest color and black the smallest.

Red, yellow and orange as warm colors are classified as they are usually associated with fire and sun. Warm colors are stimulating, aggressive and lively. Red is associated with matters of the heart—valentine, love and romance; exciting, fiery, and dangerous. It is a popular color for women's wear and it is one of the colors that are frequently used for clothing every season. Yellow is bright, sunny, cheerful, friendly and optimistic. Orange has become popular in the youth market.

Blue, green and violet remind people of the sky and the sea. Blue is quiet, restful and reserved. Denim and navy blue have become wardrobe classics. For this reason, most manufacturers include blue especially in their spring or summer lines. Green is a soothing color, suggesting peace and calm. It is used primarily in a dark value in fall sportswear lines. Violet, historically associated with royalty, has come to represent wealth, dignity and drama.

Historically, colors have been used to denote rank and profession. Golden yellow, for instance, was associated with royalty and in some periods it could be worn only by Chinese Emperor. Black became customary for the apparel of the clergy and for members of the judiciary in the West. Color symbolism often varies with geographical location. While white is the Western world's symbol of purity, worn by brides and used in communion dresses, it is the color of mourning in India and China.

Today, fashion colors always change as often as fashion itself. A new season's colors generally grow out of the existing color palette. Fashion color predictions and forecasting companies are set out well before the designers start their collections. These companies develop color stories from many sources including international fabric fairs where yarn technologists, cosmetic suppliers, trimming merchants and other related industries show their latest development on new color stories derived from existing popular colors.

The textile industry continually develops fabric technology in order to create fabrics with innovative weave, knits, texture and finishes. It is important to consider color along with texture as the surface of the fabric as much as the color presented.

4. Types of fabrics

Deciding which type of fabric to make an item with is an important decision, as fabrics can have countless qualities. From natural to synthetic fibres and from knit to woven, here's a look at different fabric types and how to identify them.

Cashmere. Cashmere is a type of wool fabric that is made from cashmere goats and pashmina goats. Cashmere is a natural fibre known for its extremely soft feel and great insulation. The fibres are very fine and delicate, feeling almost like a silk fabric to the touch. Cashmere is significantly warmer and lighter than sheep's wool. Often cashmere is made into a wool blend and mixed with other types of wool, like merino, to give it added weight, since cashmere fibres are very fine and thin.

Chiffon. Chiffon is a lightweight, plain-woven fabric with a slight shine. Chiffon has small **puckers** that make the fabric a little rough to the touch. These puckers are created through the use of s-twist and z-twist crepe yarns, which are twisted counter-clockwise and clockwise respectively. Crepe yarns are also twisted much tighter than standard yarns. The yarns are then woven in a plain weave, which means a single weft thread alternates over and under a single warp thread. The sheer fabric can be woven from a variety of textile types, both synthetic and natural, like silk, nylon, **rayon**, or **polyester**.

Gingham. Gingham is a cotton fabric, or sometimes a cotton blend fabric, made with dyed yarn woven using a plain weave to form a checked pattern. Gingham is usually a two-color pattern, and popular combinations are red and white gingham or blue and white gingham. The checked pattern can come in a variety of sizes. Gingham pattern is reversible and appears the same on both sides. Gingham is a popular fabric due to its low cost and ease of production. Gingham is used frequently for button-down shirts, dresses, and tablecloths.

Jersey. Jersey is a soft stretchy, knit fabric that was originally made from wool. Today, jersey is also made from cotton, cotton blends, and synthetic fibres. The right side of jersey knit fabric is smooth with a slight single rib knit, while the backside of jersey is piled with loops. The fabric is usually light-to-medium weight and is used for a variety of clothing and household items, like sweatshirts or bed sheets.

Lace. Lace is a delicate fabric made from yarn or thread, characterized by open-weave designs and patterns created through a variety of different methods. Lace fabric was originally made from silk and linen, but today cotton thread and synthetic fibres are both used. Lace is a decorative fabric used to accent and embellish clothing and home decor items. Lace is traditionally considered a luxury textile, as it takes a lot of time and expertise to make.

Linen. Linen is an extremely strong, lightweight fabric made from the **flax plant**. Linen is a common material used for towels, tablecloths, napkins, and bed sheets, and the term "linens", i. e. bed linens, still refers to these household items, though they are not always made out of linen fabric. The material is also used for the inner layer of jackets, hence the name "lining". It's an incredibly absorbent and breathable fabric, which makes it ideal for summer clothing, as the lightweight qualities allow air to pass through and moderate the body temperature.

Modal. Modal fabric is a semi-synthetic fabric made from beech tree pulp that is used primarily for clothing, such as underwear and pajamas, and household items, like bed sheets and towels. Modal is a form of rayon, another plant-based textile, though it is slightly more durable and flexible than rayon. Modal is often blended with other fibres like cotton and **spandex** for added strength. Modal is considered a luxurious textile thanks to both its soft feel and high cost, as it is more expensive than either cotton or **viscose**.

Polyester. Polyester is a man-made synthetic fibre created from petrochemicals, like coal and petroleum. Polyester fabric is characterized by its durable nature; however it is not breathable and doesn't absorb liquids, like sweat, well. Polyester blends are also very popular as the durable fibre can add strength to another fabric, while the other fabric makes polyester more breathable.

Words:

fleece [fli:s]	n. 羊毛，绒头织物
yarn [jɑ:n]	n. 纱；纱线
finishes ['fɪnɪʃɪs]	n. 成品 (finish 的复数)

drape [dreɪp]	n. 布帘；褶皱
vinyl ['vaɪnl]	n. 乙烯基(化学)
satin ['sætɪn]	n. 缎子
smocking ['smɒkɪŋ]	n. 装饰用衣褶
silhouette [ˌsɪlu'et]	n. 轮廓，剪影
drapery ['dreɪpəri]	n. 布料；帏帐；打褶的帐幔
chiffon ['ʃɪfɒn]	n. 薄绸；雪纺绸
motif [məʊ'tiːf]	n. 装饰图案，式样
gingham ['gɪŋəm]	n. 条纹棉布；条格平布
twill [twɪl]	n. 斜纹布；斜纹织品
chambray ['ʃæmbreɪ]	n. 有条纹或格子花纹的布
linen ['lɪnɪn]	n. 亚麻布；亚麻制品
dimity ['dɪmɪtɪ]	n. 凸花条纹布
paisley ['peɪzli]	n. 佩斯利(旋涡纹)图案
acetate ['æsɪteɪt]	n. 醋酸纤维素及其制成的产品
jersey ['dʒɜːzi]	n. 运动衫，毛线衫
crepe [kreɪp]	n. 绉纱；绉绸
clinch [klɪntʃ]	v. 确定，敲定，解决
cashmere ['kæʃmɪə(r)]	n. 开司米；山羊绒；克什米尔羊毛
pucker ['pʌkə(r)]	n. 皱纹；皱褶
rayon ['reɪɒn]	n. 人造丝；人造纤维
polyester [ˌpɒli'estə(r)]	n. 聚酯
spandex ['spændeks]	n. 氨纶，弹性纤维
viscose ['vɪskəʊz]	n. 纤维胶；人造丝

Phrases：

pin stripe	细条纹
flax plant	亚麻树

Chapter 3 Textile and Life

Critical reading and thinking

Task 1 Overview

Work in pairs and retell the differences of fabrics. Use as many lexical chunks as possible.

Task 2 Group discussion

Work in groups of 4-5 and have a discussion about the following questions.
1. How do people identify the different textures of different kinds of clothes?
2. Do colors play an important role in deciding the buying of clothes? Why?

Task 3 Language building-up

Ⅰ. Translate the following terms from English into Chinese or vice versa.

strongest visual impact	
lightweight fabric	
breathable fabric	
single warp thread	
jersey knit fabric	
平纹机织物	
羊毛混纺	
印染织物	
机织染色纱	
Z 捻绉纱	

Ⅱ. Complete the following sentences with the words or phrases given in the previous exercise.

1. The fancy design with bright colors and exaggerated motifs can produce the _____ on the audience.
2. The masks, which are used to prevent the spread of the coronavirus, are comfortable because they're made with _____, according to the maker.

3. On textiles, Rwanda can supply _____ of cotton, while Burundi can export sacks and bags for packing of goods, worn clothing and clothing accessories.
4. These puckers are created through the use of s-twist and _____, which are twisted counter-clockwise and clockwise respectively.
5. _____ convey new fashion trends easily and comparatively quickly.

Task 4 Translation

I. Translate the following paragraph into Chinese.

Synthetic fibres are man-made fibres. Most of the synthetic fibres are made from polymers produced by polymerization. Synthetic fibres are manufactured usually from oil, coal or natural gas. Synthetic fabrics are the most prevalent in the world. China is the largest producer accounting for 70% of total global production. India is the second largest producer of synthetic fibre, but only 7.64% of global production comes from India while the European Union is the largest importer of synthetic filament fibres. The EU is followed by Turkey and the United States. Within the European Union, Germany and Italy are among the largest importers. There are many other importing countries such as the Middle East and African countries.

II. Translate the following paragraph into English.

随着如今经济全球化的发展，中西服饰文化融合趋势也空前加强。中国服装界正在努力同世界接轨，走一条时尚加民族特色道路，在传统服饰设计中融入西方时尚元素，同时中国元素正在影响着国际时装界的发展。我们在探讨中西服饰文化差异的同时，更应该思考在全球化的冲击下保持民族特色的重要性。关键在于找到这个点，找到民族文化和大同文化切入的融合点。

Task 5 Research

Surf the Internet for more information about the textile industry.
(You can try http://www.historyofclothing.com/textile-history/, https://www.thoughtco.com/the-history-of-textiles-172909.)

Task 6 Further reading

Passage 1

Quick Guide to 2000s Fashion

We may be almost two decades removed from the dawn of the new millennium, but it can be weird to think of anything set in the 2000s as a period piece. From movies set entirely in the decade, like 2018's *Lady Bird*, or shows that constantly feature flashbacks, like *This Is Us*, many productions are exploring the recent past. So if you don't know a lot about the aughts or have just forgotten, here's what you need to know about 2000s fashion to put together the perfect look.

What '00s look are you going for?

Many clothing trends in the 2000s were born out of globalization, the rise of fast fashion (affordable clothes based off runway designs usually found in department stores like Mervyn's, J. C. Penney, and Macy's), and celebrities' growing influence as style icons.

As with all decades, looks and trends varied as the '00s went on, so if you're tasked with putting an outfit together, don't assume your tracksuit straight from 2001 will work for every occasion. The 2000s were also home to trends that hit big and burnt out fast. If you're going for a particular look, be sure that style is from the right time period before heading to set.

Early 2000s fashion

Technology and Y2K had a huge impact on fashion in the early 2000s. The color palette was filled with shiny black tones and reflective metallics. While some Y2K trends were worn daily, many of these looks were reserved for going out. Popular outfits for women included mesh or handkerchief tops, box-pleated or leather skirts, shiny pants, and sparkly shoes. For men, Y2K looks usually involved leather jackets, a statement dress shirt, and chunky shoes. If you need inspiration, Britney Spears and *NSYNC were trendsetters for this type of style.

Casual clothing and leisurewear were the other big trends of the early 2000s. Denim became a staple for men and women, going beyond jeans to shirts, jackets, and hats. If you're putting together a casual style for women, think crop tops, hoodies, low-rise flared jeans, cargo pants, daisy dukes, jean skirts, off the shoulder tops, ribbed sweaters, with platform sandals, UGG boots, or sneakers. Common looks for men included distressed denim, cargo pants, tracksuits, rugby or polo shirts, flip flops, oxford shoes, and sneakers.

Mid 2000s fashion

As the decade went on, 2000s fashion began to take cues from 1960s bohemian looks. Yoga pants, low-rise jeans, cowl-neck shirts, peasant tops, capri pants, cropped jackets, and dresses over jeans were popular outfit choices for women. These were often paired with accessories like chunky belts, aviator sunglasses, jelly bracelets, ties worn around the neck or as belts, ballet flats, and platform boots.

The 1960s revival looks were also popular with men. Outfit options included light wash bootcut jeans, cargo shorts, classic rock t-shirts, fitted cowboy shirts, henleys, polos with popped collars, and seersucker suits.

The middle of the decade was also defined by the latest "it" items, like Von Dutch trucker hats, Juicy Couture tracksuits, and Louis Vuitton bags.

Late 2000s fashion

Many styles from the early and mid-2000s carried over to the latter part of the decade, with a few exceptions. For women, crop tops were replaced with camisoles and miniskirts gave way to baby doll, bubble skirt, and sweater dresses. There was also a 1980s and 1990s revival that reintroduced neon colors, animal prints, geometric shapes, light denim jeggings, and ripped acid washed jeans that were worn with gladiator sandals, ballet flats, and headbands. An oversized look started to gain popularity, but it was subtler than the traditional '80s fit.

Men's late 2000s fashion was a mix of 1950s and 1980s throwbacks, with letterman and black leather jackets, overcoats, slim cut jeans, Ed Hardy t-shirts, flannel shirts, and V-neck sweaters. These were often paired with dad hats, wayfarers or aviators, motorcycle boots, Converse, Vans, or sneakers. The men's power suit was also updated from the '80s to have more a slim tailored cut.

Types of '00s looks
Hip-Hop

Many rappers influenced fashion with their own clothing lines in the 2000s, including Jay-Z, Diddy, Nelly, and 50 Cent. Many looks included baggy jeans, tall t-shirts, sports jerseys, velour suits, bubble jackets, and puffer vests paired with headbands, sweatbands, Timberland boots, and sneakers like Adidas Superstars or Nike Air Force 1s.

Streetwear

Streetwear is often brand focused casual clothing, like jeans, tees, and sneakers. In the late 2000s, popular streetwear styles included distressed skinny jeans, loose fitting tops, loose or fitted tracksuits, track pants, hoodies, graphic t-shirts, vintage thrift shop tees, and Tommy Hilfiger and U. S. Polo Assn. brands. Sneakers were an important part of the look, especially retro Nike Air Jordan's and Adidas Yeezys. Shoulder bags were also a popular accessory for men.

Emo

Emo fashion trickled into the mainstream in the mid-2000s and took cues from goth and punk styles. Outfits were often all or mostly black with skinny jeans, band t-shirts, studded belts, and checkered Vans. No emo outfit was complete without the right hairstyle, most common were choppy cuts with long side-swept bangs dyed black, platinum blonde, or a bright color.

Scene

Scene styles became popular at the end of the decade and were influenced by indie, rave, and punk fashion. This look often included skinny black biker jeans, bright t-shirts and hoodies, band shirts, tutus, and studded belts. Like with emo styles, hair was a big part of the look. A popular scene hairstyle was bright neon dyed hair teased and swept to the side and accessorized with tiaras and bows.

Leisurewear

If you wanted to be comfortable yet fashionable in the early 2000s, all you had to do was throw on your favorite tracksuit. Everyone from Britney Spears to Beyoncé and Eminem

to Diddy were in on the tracksuit craze. They were often brightly colored and emblazoned with rhinestone logos and phrases. It was also popular to mix track pants with dress clothing and designer shoes to elevate the look.

Passage 2

Fashion Moves in a Circle

Fashion is a reflection of wealth and prestige can be used to explain the popularity of many styles throughout clothing history. While everything repeats itself in life, fashion is no exception. What we now call "novelties" and "fashion trends" were all invented long ago and have already been tried by fashionistas.

When we decided to find out where popular fashion trends came from, we again realized that most of them have already been invented and followed by fashion lovers over the years.

Trenches from the 1940s

1901 was the year when the word "trench" first appeared in the fashion world, when the British military ministry approved it as clothing for soldiers. After World War I, the trench was transferred to the cloakrooms of civilians, and fashionistas all over the world fell in love with it. A huge role in this was played by the movies of the '40s-'60s: The trench appeared in *Casablanca*, *The Key*, *Breakfast at Tiffany's*, and many other films. It has become a truly cult item.

Nighties from the 1990s

This slip dress looks like a real nightie, as if you've just woken up and immediately decided to go to a party. It remains a favorite outfit of fashion lovers, both today and 2 decades ago.

Chokers from the 2000s

Probably each and every girl of the noughties had a trendy collar called a "choker". Today celebrities again set us an example of being stylish by wearing this kind of accessory.

Frills from the 19th century

Frills, jabots—all this cute girly stuff has once again returned to us.

They were popular in the 1920s and the 2000s and were especially popular among

fashionable ladies of the previous century.
Bell-bottoms from the 2000s

Bell-bottoms (or flares) have gained extreme popularity again. But now fashionistas combine flares with long pants or culottes.

Peplum from the 1940s

An overskirt, or peplum, perfectly emphasizes the beauty of a woman's shape, at the same time hiding any imperfections. This indispensable piece of clothing is now mostly worn with tight jeans or skirts, but, of course, you are free to experiment and set new trends in fashion.

Midi skirts from the 1930s

Mini and maxi skirts have been put off until better times. The absolute favorite of modern fashion is a midi skirt. Though even this isn't a novelty—it was popular 80 years ago.

Cuffed jeans from the 1990s

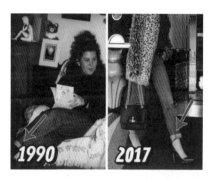

In the 1990s, we made small cuffs on jeans called "pyramids". Shortened pants are on trend today, and you can make your casual jeans more fashionable by making wide cuffs.

"Banana jeans" or "pyramids" from the 1980s

Tapered jeans—also called "banana jeans" or "pyramids"—were extremely popular in the '90s among both girls and boys. In 2017 they are popular again, but now they suit girls only.

Socks with shoes or sandals from the 1940s

Socks with shoes or sandals were a fashion trend in the 1930s and 1950s. Then the excitement subsided, and this funny combination was forgotten for a while. But in 2014, socks with shoes and sandals suddenly reappeared in the fashion industry. This trend later moved from fashion shows to the everyday life of true fashionistas. However, not all fashion experts agree that this combination is successful. So be careful!

Oversized coats from the 1980s

The oversized coat that looks like your man just put his overcoat on your shoulders so you won't get cold on a chilly night. It's back!

Pointed toe shoes from the 2000s

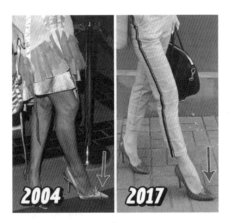

Shoes with very pointed toes were an iconic item for fashionistas in the noughties. In 2017, pointed shoes are trendy again. They're more elegant and delicate but still stylish.

Culottes from the 1950s

Culottes have come a long way in fashion and have often been an indispensable trendy item of ladies' clothing. Today we love and wear them again. They can be part of a casual look or worn as part of a formal outfit. Try different lengths, colors, and materials.

Swimsuits from the 1950s

Swimsuits as we know them today appeared in the 1950s and immediately became popular among ladies. Since then a lot of models, forms, and styles have changed, from conservative swimming costumes up to ultra hot bikinis. Now, more than 60 years later, one-piece swimsuits are extremely trendy again.

Chapter 4 Textile and Manufacture

Section 1 Textile Fibres

Lead-in

Warm-up questions:

1. What is textile fibre? Can you name some commonly used textile fibres? What are natural fibres and what are man-made fibres?
   (1) natural fibres _____ _____ _____
   (2) man-made fibres _____ _____ _____
2. Can you identify the following pictures of the cross-sectional view and longitudinal view of textile fibres under microscope? Please match the pictures A-D with the corresponding fibres 1-4.
   (   ) (1) Wool   (   ) (2) Polyester   (   ) (3) Cotton   (   ) (4) Silk

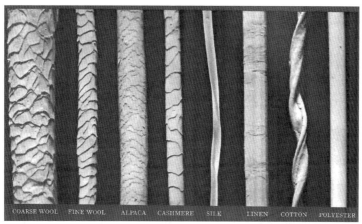

E

Lexical chunk bank	
纺织纤维	textile fibre
动物纤维	animal fibre
植物纤维	plant/vegetable fibre
矿物纤维	mineral fibre
再生纤维	regenerated fibre
合成纤维	synthetic fibre
纤维素纤维	cellulose fibre
蛋白质纤维	protein fibre
短纤维	staple fibre
长丝纤维	filament fibre

1. Introduction of textile fibres

What is fibre?

Generally speaking, materials with diametre ranging from several **micrometres** to tens or hundreds of micrometres and with length at least 100 times their diameter can be considered to be fibres.

What is textile fibre?

The term "**textile fibre**" means a unit of matter which is capable of being **spun** into a yarn, a **cord** or made into a fabric, is the basic structural element of textile products. Not all fibres are suitable for textiles. Textile fibres must have certain essential properties if they are to qualify as suitable substances for use in forming textile fibres. These primary properties include a high **length-to-width ratio**, **tenacity** or adequate strength, **flexibility** or **pliability**, **cohesiveness** or spinning quality, and **uniformity**. Secondary properties include physical shape, density, luster, **moisture regain**, **elastic recovery**, **elongation**, **resiliency**, thermal behavior, resistance to **biological organisms**, and resistance to chemical and other environmental conditions. In other words, this means they need to be strong enough to hold their shape, flexible enough to be shaped into a fabric or yarn, elastic enough to stretch, and durable enough to last. Textile fibres also have to be a minimum of 5 millimetres in length. Shorter fibres cannot be spun together.

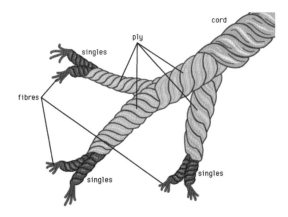

Textile fibres may be **staple** or **filament**. **Staple fibres** are relatively short, measured in millimetres or inches. **Filament fibres** are relatively long, measured in metres or yards. Most natural fibres are staple fibers. The only natural filament fibre is **reeled or cultivated silk**. On the other hand, all man-made fibres can be staple or filament; they begin as filament, and in this form can make silky or silk-like fabrics. They can also be cut or broken into staple to make fabrics that look and feel more like wool, cotton, or flax/linen.

Textile fibres are normally classified as either natural fibres or man-made fibres.

2. Natural fibres

All fibres which come from natural sources (animals, plants, mineral, etc.) and do not require fibre formation or reformation are classified as natural fibres. Natural fibres include the **protein fibres** or **animal fibres** such as wool and silk, the **cellulose fibres** or **plant/vegetable fibres** such as cotton and linen, and the **mineral fibres such as asbestos**.

Natural fibres with their long history of serving mankind are very important in a wide range of applications, and they compete and co-exist in the 21st century with man-made fibres, especially as far as quality, **sustainability** and economy of production are concerned. Natural fibres conduct heat, can be properly dyed, resist **mildew**, have natural **antibacterial properties, block UV radiation and can be easily made flame retardant**. Genetic modification of natural **fibrous** raw materials improves their productivity and performance.

The following is a classification of common natural fibres.

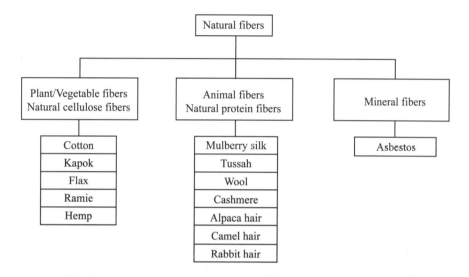

Cotton

The word "cotton" comes from the Arabic word "quton". The earliest production of cotton was in India, where the material dates back to 5000 BC.

Cotton, the seed hair of plants of the **genus** Gossypium(棉属，棉花), and the purest form of cellulose available in nature, is the dominant natural fibre. Cotton is a good conductor of heat, is **susceptible** to damage by mildew, turns yellow and becomes weak when exposed to prolonged sunlight. It is easily **flammable** and has rather poor resistance to wear. The highest quality cotton varieties have the longest fibres, thin, with good **resilience** and elasticity which makes them easy to spin, suitable for the production of high-quality goods—especially garments. Recent advances have included production of **genetically modified cotton (GM)** and also **organic cotton**.

Cotton has a number of distinguishing characteristics that make it such a popular fibre in the textile industry: softness, durability, **absorbency**, and **breathability**.

There are four different types of cotton, each with its own characteristics: **Sea Island Cotton, Egyptian Cotton, Upland Cotton, Organic Cotton.**

Flax

Flax (Linen) is the oldest natural fibre used by our ancestors as early as 10,000-8,000 BC, when they changed their way of life from **nomadic** hunting and gathering to a more **sedentary**, agrarian style of living. Until the 18th century, flax was the dominant fibre in Europe, but later on cotton, cultivated in America and India, which began to systematically replace flax.

Flax is valued for its strength, lustre, durability, and **moisture absorbency**. Linen is a bast fibre obtained from the stem of the flax plant. It is resistant to attack by **microorganisms**, and its smooth surface repels dirt. It is stronger than cotton, dries more quickly, and is more slowly affected by exposure to sunlight. Low elasticity, imparting hard, smooth texture, also makes flax subject to wrinkling, which can be reduced by chemical treatment. Because flax absorbs and releases moisture quickly and is a good conductor of heat, flax garments have a cooling effect on the wearer.

Wool

Wool was probably the first animal fibre to be made into cloth. The art of spinning wool into yarn developed about 4000 BC and encouraged trade among the nations in the region of the Mediterranean Sea.

Wool is a natural fibre also known as fleece, similar to human hair or fur, that covers the skin of a variety of animals including sheep, goats, and **alpacas**, and is used in different **woven and knitted textiles.** The fibres are collected during the annual **shearing** of the animals, and then processed. Wool has been used for thousands of years to make warm clothing, blankets, and other furnishings; regardless of which animal produces them, the natural **crimp** of the wool fibres makes them an effective **insulator** against the cold which also increases the elasticity and elongation properties of the fibres. The quality and other properties of the wool, however, vary considerably, depending on the type and breed of animal.

Cashmere

Cashmere comes from goats and is softer still than sheep's wool. Specifically, it is produced from fibres that make up the soft, downy undercoat of **Kashmir** goats that originally inhabited in China, India, Iran, Afghanistan, and Iraq. It is one of the rarest specialty fibres in the world. Due to its high-value, it is also called "diamond fibre" and "soft gold fibre".

Goats grow hairs for the cold winter to keep themselves warm. In the spring and summer, goats' hairs naturally fall off. Goats are combed with a special comb to collect their hairs. These thin and long hairs are manufactured into delicate and warm cashmere wears. Collecting these particular hairs is a laborious task that must be done by hand during spring **molting** season, resulting in a relatively small yield. The difficulty involved with collecting cashmere wool explains why it is relatively rare and considered a luxury item compared to sheep's wool. It takes at least two goats to make one **two-ply** cashmere sweater, whereas the wool from one sheep can be used to make four or five conventional wool sweaters. Cashmere wool is softer and lighter than sheep's wool and possesses a higher **loft**, which translates into **plush**, luxurious fabrics with a silky feel. Because it also **drapes** beautifully, cashmere garments impart an elegant look on more formal occasions.

 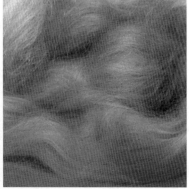

Silk

Silk is one of the oldest fibres known by humankind, which has been used in textile manufacture for at least 5,000 years. From its discovery, in about 2650 BC, until today, silk has been known as the "queen of the fibres". China and Japan are the world's major sources of silk, although minor quantities come from India, Italy, and Korea.

The two main categories are **cultivated silk** (**mulberry silk**) **and wild silk** (**tussah**). Cultivated silk is more lustrous and lighter in color than tussah.

Silk is a natural fibre produced by insects as a material for their nests and cocoons. Made primarily of a protein called **fibroin**, silk is known for its shine and softness as a material. Silk is marked by the following properties: low density for light and comfortable clothing, high resistance to deformation and good insulation, keeping warm in winter and cool in summer.

3. Man-made fibres

Man-made fibres are fibres in which either the basic chemical units have been formed by **chemical synthesis** followed by fibre formation, or the **polymers** from natural sources have been dissolved and regenerated after passage through a **spinneret** to form fibres. Man made fibres can be classified to be **organic and inorganic fibres**. Organic fibres can be sub-classified into two types: **regenerated fibres** and **synthetic fibres**.

Regenerated fibres are similar to cotton. They were the first of the manufactured fibres to be developed. They are made from cellulose-based fibres that originate from plants such as wood **pulp**. A chemical is added to extract the cellulose fibres. The classification of the fibres relates to the **chemical solvent** system used to extract the fibres, so regenerated fibres are part natural and part artificial. Viscose(CV), Cupro(CUP)(铜氨纤维), **acetate**(**CA**), **triacetate**(**CTA**), **Modal**(**CMD**) and Lyocell(CLY)(绿赛尔纤维) or Tencel(天丝) are all regenerated cellulose fibres. Both Viscose and Cupro are called **rayon**. As regenerated fibres come from plant-based sources, their properties are similar to those of cotton. They are highly absorbent, washable, soft, smooth, comfortable to wear and have good drape. Owing to their properties, regenerated fibres are widely used in clothing. They can be given different

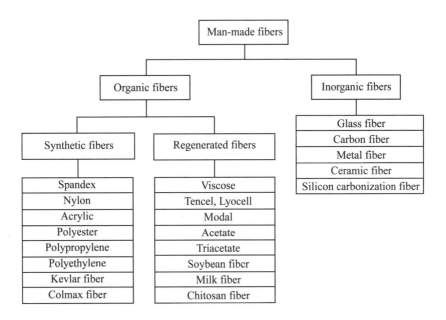

finishes to make them smooth, shiny or textured. Regenerated protein fibres involve **soybean fibres, milk fibres and chitosan fibres**, which are particularly used for medical applications.

Another group of man-made fibres (and by far the larger group) is the synthetic fibres. Synthetic fibres are made of polymers that do not occur naturally but instead are produced entirely in the chemical plant or laboratory, almost always from by-products of petroleum, coal or natural gas. The synthetic fibres include **polyethylene (PE), polyamide (PA) or Nylon, polyester (PES)** or **Dacron, polyacrylonitrile (PAN)** or **acrylic, polypropylene (PP), polyurethane (PU)** or spandex, and so on. Synthetic fibres can be mass-produced to almost any set of required properties.

Advantages of synthetic fibres
- Synthetic fibres are very durable and do not wrinkle easily.
- They are elastic and can be easily stretched out.
- They are strong and can sustain the heavy load.
- They are soft and hence used in clothing material.
- They are cheaper as compared to natural fibres.

Disadvantages of synthetic fibres
- Most synthetic fibres do not absorb moisture.
- Synthetic fibres do not easily take up colours.
- They can be affected if washed using hot water.
- They tend to catch fire, burn faster and melt easily.

Inorganic fibres are the fibres made from inorganic materials and usually include **carbon**

fibres, **ceramic fibres**, **glass fibres**, **silicon carbonization fibres** and **metal fibres**. They are mostly used for some special purposes in order to perform some special functions.

Words:

cross-sectional [ˈkrɒːsˈsekʃnəl]	*adj.* 截面的，断面的，剖面的
longitudinal [ˌlɒŋgɪˈtjuːdɪnl]	*adj.* 经度的；纵向的；纵的；纵观的
micrometre [maɪˈkrɒmɪtə]	*n.* 微米
spin [spɪn]	*n. /v.* 纺纱；纺丝
cord [kɔːd]	*n.* 绳
tenacity [təˈnæsətɪ]	*n.* 韧性；柔韧性
pliability [ˌplaɪəˈbɪlɪtɪ]	*n.* 柔韧性；可弯性
flexibility [ˌfleksəˈbɪlətɪ]	*n.* 柔韧性
cohesiveness [kəʊˈhiːsɪvnəs]	*n.* 抱合力；黏结性，内聚性
uniformity [ˌjuːnɪˈfɔːmətɪ]	*n.* 均匀性
elongation [ˌiːlɒŋˈgeɪʃn]	*n.* 伸长；伸长率
resiliency [rɪˈzɪlɪənsɪ]	*n.* 弹性；弹力
staple [ˈsteɪpl]	*n.* 纤维，短纤维
filament [ˈfɪləmənt]	*n.* 长丝
cellulose [ˈseljuləʊs]	*n.* 纤维素
asbestos [æsˈbestəs]	*n.* 石棉
sustainability [səˌsteɪnəˈbɪlətɪ]	*n.* 持续性，能维持性
mildew [ˈmɪldjuː]	*n.* 霉
antibacterial [ˌæntibækˈtɪəriəl]	*adj.* 抗菌的
fibrous [ˈfaɪbrəs]	*adj.* 含纤维的，纤维性的
genus [ˈdʒiːnəs]	*n.* (动植物的)属；类；种；型
susceptible [səˈseptəbl]	*adj.* 易受影响的；易受感染的
flammable [ˈflæməbl]	*adj.* 易燃的，可燃的
resilience [rɪˈzɪlɪəns]	*n.* 弹性；弹力
absorbency [əbˈzɔːbənsɪ]	*n.* 吸收性

breathability [breθə'bɪlɪtɪ]	n. 透气性
nomadic [nəʊ'mædɪk]	adj. 游牧的；流浪的
sedentary ['sedntri]	adj. 坐着的；(指人)不爱活动的
microorganism [ˌmaɪkrəʊ'ɔːgənɪzəm]	n. 微生物
alpaca [æl'pækə]	n. (南美的)羊驼，羊驼毛，羊驼呢(织物)
shearing ['ʃɪərɪŋ]	n. 剪羊毛，剪取的羊毛
crimp [krɪmp]	n. 卷曲
insulator ['ɪnsjuleɪtə(r)]	n. 绝缘、隔热或隔音等的物质或装置
Kashmir ['kæʃmɪə]	n. 克什米尔
molting ['məʊltɪŋ]	n. 换羽，脱毛；蜕皮
two-ply ['tuː'plaɪ]	adj. 双股的
loft [lɒft]	n. 弹性；蓬松
plush [plʌʃ]	n. 长毛绒；长绒棉
drape [dreɪp]	v. 悬垂
tussah [tʌ'sɑː]	n. 柞蚕丝；柞蚕，野蚕丝
fibroin ['faɪbrəʊɪn]	n. 蚕丝蛋白
synthesis ['sɪnθəsɪs]	n. <化>合成
polymer ['pɒlɪmə(r)]	n. [高分子] 聚合物；多聚物
spinneret ['spɪnəˌret]	n. 喷丝头；喷丝板
regenerated [rɪ'dʒɛnərɪt]	adj. 再生的
pulp [pʌlp]	n. 木浆，黏浆状物质
acetate ['æsɪteɪt]	n. 醋酸盐；醋酸酯
triacetate [traɪ'æsɪteɪt]	n. 三醋酸酯，三乙酸酯
Modal ['məʊdəl]	n. 莫代尔纤维
rayon ['reɪɒn]	n. 人造丝，人造纤维丝
finish ['fɪnɪʃ]	n. 整理
polyethylene [ˌpɒli'eθəliːn]	n. 聚乙烯
polyamide [ˌpɒlɪ'æmaɪd]	n. 聚酰胺

polyacrylonitrile [ˌpɒlɪˈækrələʊˈnaɪtrɪl]	n. 聚丙烯腈
polypropylene [ˌpɒliˈprəʊpəliːn]	n. 聚丙烯
polyurethane [ˌpɒliˈjʊərəθeɪn]	n. 聚氨酯
nylon [ˈnaɪlɒn]	n. 尼龙；尼龙织品
polyester [ˌpɒliˈestə(r)]	n. 聚酯
dacron [ˈdekrɑn]	n. 涤纶；的确良
acrylic [əˈkrɪlɪk]	n. 腈纶；聚丙烯腈
carbonization [ˌkɑːbənaɪˈzeɪʃn]	n. 碳化

Phrases：

length-to-width ratio	长径比
moisture regain	回潮率
elastic recovery	回弹性
biological organism	生物有机体
reeled silk	绞丝
cultivated silk	家蚕丝；桑蚕丝
flame retardant	阻燃的
genetically modified (GM) cotton	转基因棉
Sea Island Cotton	海岛棉
Egyptian Cotton	埃及棉
Upland Cotton	陆地棉
Organic Cotton	有机棉
moisture absorbency	吸湿性
woven textile	机织纺织品
knitted textile	针织纺织品
mulberry silk	桑蚕丝
wild silk	野蚕丝；柞蚕丝
organic fibre	有机纤维

inorganic fibre	无机纤维
chemical synthesis	化学合成
chemical solvent	化学溶剂
soybean fibre	大豆纤维
milk fibre	牛奶纤维
chitosan fibre	甲壳素纤维
carbon fibre	碳纤维
ceramic fibre	陶瓷纤维
glass fibre	玻璃纤维
silicon carbonization fibre	碳化硅纤维
metal fibre	金属纤维

Critical reading and thinking

Task 1 Overview

Work in pairs and retell the major properties of natural fibres such as cotton, flax, cashmere, wool and silk. Use as many lexical chunks as possible.

Natural fibres	Major properties
cotton	
flax	
cashmere	
wool	
silk	

Task 2 Group discussion

Work in groups of 4-5 and have a discussion about the following questions.

1. Which natural fibres and which man-made fibres would tend to have properties that are very similar?

2. What are the advantages and disadvantages of the synthetic fibres?

Task 3 Language building-up

I. Translate the following terms from English into Chinese or vice versa.

filament fibre	
cellulose fibre	
protein fibre	
regenerated fibre	
flame retardant	
碳纤维	
陆地棉	
桑蚕丝	
吸湿性	
短纤维	

II. Complete the following sentences with the words or phrases given in the previous exercise.

1. Flax is valued for its strength, lustre, durability, and _____.
2. _____ are relatively long, measured in metres or yards. Most natural fibres are staple fibres. The only natural filament fibre is reeled or cultivated silk.
3. _____ are similar to cotton. They were the first of the manufactured fibres to be developed. They are made from cellulose-based fibres that originate from plants such as wood pulp. A chemical is added to extract the cellulose fibres.
4. Natural fibres include the protein fibres or animal fibres such as wool and silk, the _____ or plant/vegetable fibres such as cotton and linen, and the mineral fibre—asbestos.
5. Natural fibres conduct heat, can be properly dyed, resist mildew, have natural antibacterial properties, block UV radiation and can be easily made _____.

Task 4 Guessing game

Guess the missing lexical chunks from different contexts with the ones in the box.

| A. textile fibres | B. moisture regain | C. synthetic fibres |

Group 1

1. The amount of _____ and moisture content is not constant, it keeps changing. This changes the amount of moisture in the material.
2. _____ is defined as the amount of water (or moisture) expressed as the percentage of oven-dry weight.
3. _____ by some of these materials causes them to become sticky and to not flow freely causing catastrophic problems in candy making, materials conveying, pharmaceutical and vitamin tablet making and packaging.

Group 2

1. _____ are those which have properties that allow them to be spun into yarn or directly made into fabric.
2. If two _____ are twisted together, it is stronger than both individual without increasing its tenacity.
3. _____ are generally divided into two broad categories: synthetic fibres and natural fibres.

Group 3

1. _____ are more prone to static buildup than natural fibres like cotton.
2. Like natural fibres, _____ do not easily take up colors. These fibres can burn faster than natural fibres. They are prone to damage to heat and easily melt.
3. _____ is used in the manufacture of ropes, nets for fishing and seat belts.

Task 5 Translation

Translate the following paragraphs into Chinese.

The exceptional elasticity of spandex fibres increases the clothing's pressure comfort, enhancing the ease of body movements. Pressure comfort is the response towards clothing by the human body's pressure receptors (mechanoreceptors present in skin sensory cell). The sensation response is affected mainly by the stretch, snug, loose, heavy, lightweight, soft, and stiff structure of the material.

The elasticity and strength (stretching up to five-times its length) of spandex has been incorporated into a wide range of garments, especially in skin-tight garments. A benefit of spandex is its significant strength and elasticity and its ability to return to the original shape after stretching and faster drying than ordinary fabrics. For clothing, spandex is usually mixed with cotton or polyester, and accounts for a small percentage of the final fabric, which therefore retains most of the look and feel of the other fibres. In North America it is rare in men's clothing, but prevalent in women's. An estimated 80% of clothing sold in the United States contained spandex in 2010.

Task 6 Research

Surf the Internet for more information about textile fibres and their properties.
(You can try https://www.textileschool.com/3026/textile-fabric-types-by-fibre-sources/, https://study.com/academy/lesson/textile-fibres-definition-properties-types.html.)

Section 2　Yarns

Lead-in

Warm-up questions:

1. What is spinning? Do you know anything about spinning?
2. Have you ever seen the pictures below? Which is the traditional spinning and which is the modern spinning?

3. What does "S" and "Z of a yarn symbolize respectively as shown in the following picture?

Lexical chunk bank	
短纤纱	spun yarn/staple yarn
长丝纱	filament yarn/thrown yarn
熔体纺丝	melt spinning
溶液纺丝	solution spinning

干法纺丝	dry spinning
湿法纺丝	wet spinning
环锭纺	ring spinning
普梳纱	combed yarn
精梳纱	carded yarn
定长制	weight based system/fixed length system
定重制	length based system/fixed weight system

1. Introduction of spinning

Yarns can be produced from staple fibres or continuous filament fibres. Yarns composed of short, staple fibres may be called **spun yarns** or **staple yarns**. Yarns made from long, filament fibres maybe identified as **filament yarns** or **thrown yarns**.

Spinning is the process of taking textile fibres and filaments and making them into yarn. For thousands of years, people spun natural fibres into yarn by hand. Today, spinning involves many methods and different machines, depending on what kind of yarn is being made. Before we discuss how spinning works, let's review some basics. Fibres are short, natural hairs that come from plants like cotton and animals like sheep. Filaments are long continuous single **strands**. Silk is a natural filament, but most filaments are synthetic or man-made materials, like polyester and nylon. Converting fibres and the substances that form synthetic filaments into yarn involves different methods of spinning.

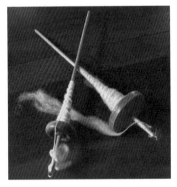 

First, let's look at the basics of spinning, which can be done with very simple tools. To make a natural yarn, you gather cleaned and prepared natural fibres like wool or cotton. Using a tool called a spindle, a rounded stick with tapered(圆锥形的) ends, you pull the fibres by hand onto it as the spindle twists. To aid the process, the spindle is weighted by something called a whorl, a small, round stone or piece of wood which allows the spinning to be maintained at a regular speed. The process of pulling and twisting results in a piece of yarn.

2. Spinning synthetic filaments

The processes used to turn synthetic substances into filaments are a bit different because they start with substances that are not yet in strand-like form. Let's look at two common methods of spinning synthetic filaments. Man-made fibres are manufactured by spinning the polymer. There are two major types of spinning process: solution spinning and melt spinning.

2.1 Solution spinning

Solution spinning is used when the desired polymer does not form a stable melt. These polymers are dissolved in a **solution** to liquefy(溶解) them rather than melting them. The two main types of solution spinning are dry and wet spinning.

2.1.1 Wet spinning

In wet spinning the polymer is dissolved in a non-volatile solvent (see Fig. a). The spinneret is located in a coagulating bath(凝固浴) that causes the fibre to precipitate(沉淀) out. The liquid in the coagulating bath is chosen so that the solvent is readily **soluble** in the liquid but the polymer is not. This will cause the polymer to precipitate out and form the desired solid filaments. The solvent can later be recovered by treating the waste water in the bath. The drag on the filament by the viscous(黏性的) liquid in the coagulating bath significantly reduces the production speed of the process. This makes wet spinning a slower process than melt or dry spinning.

The synthetic fibres produced by solution spinning are made into threads which are woven into fabric for everything from clothes and carpets to airbags and wind sails(风帆). In addition, **sterilized** synthetic fibres are used in the medical industry as **sutures**, **dressings**, and operating room drapes, face masks, caps and overshoes. Many synthetic fibres can be cut into tiny pieces called **flock**. Flock is used to make **velvet** or suede-like (像绒面的) materials or added to plastics or papers for a decorative finish.

2.1.2 Dry spinning

In dry spinning the polymer is dissolved in a **volatile solvent** (see Fig. b). Once dissolved, the polymer solution is extruded through the spinneret which is in an enclosed drying tower. The solution goes through a drying process in this tower, where the solvent is evaporated. The spinning solution is often **filtered** many times to remove foreign materials that might cause thread defects or clogging(堵塞) in the machine. The solution is extruded into the drying tower through which hot air or other gases pass. The concentration of the polymer increases as the solvent evaporates, leaving a solid polymer filament behind. The solvent is later recovered from the gas by **condensation** or absorption and is then recycled.

2.2 Melt spinning

Melt spinning used for polymers that can be melted easily, is the most widely used form of fibre spinning. In melt spinning, either molten polymer is used or **polymer chips** are melted down (see Fig. c). Once the filaments are extruded they are cooled in a fluid medium such as air, gas, or even water. In melt spinning, the molten or melted polymer is pushed vertically downwards through the spinneret. When the filaments emerge from the spinneret they are cooled by a gas, usually air, which flows **perpendicularly** across to the filaments.

The filaments can then be brought together and/or twisted together to form a thread. Before winding the thread on the **bobbin**, it is often treated with water or a wetting agent and then a **lubricant**. Later, the wound thread may be dyed and drawn on other machines. Spinneret design is dependent on the type of spinning and desired cross-sectional shape of the filament. The cross-sectional shape determines different features the filament will exhibit, such as how it will reflect light, the ability to **insulate**, and how easily it shows dirt.

Some common fibres produced using melt spinning are: polypropylene, polyester, and nylon. Nylon was the world's first synthetic fibre discovered and was initially used for tooth brush **bristles** before being used to produce women's stockings during the 1940's.

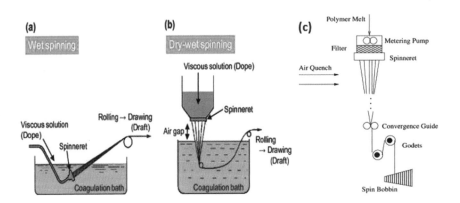

3. Spinning natural fibres

This basic spinning process was used for natural fibres. It's how the process worked for thousands of years. Then, advances in tools and technology made spinning more efficient. In 1828, a machine was invented that allowed a method called **ring spinning** to mechanize the process. During this process, hundreds of spindles are mounted vertically on a machine that spins fibres into yarns. Now take cotton yarn spinning for example, the major production processes of ring spinning are as follows.

3.1 Opening, cleaning and blending

Opening, cleaning and blending is the first operation which takes place in the spinning mill. The department where opening, cleaning and blending is performed is called **blow room**, because the cotton is transported from one machine to another by blowing it through the ducts(管道). Blow room is combination of machines varying significantly unlike other

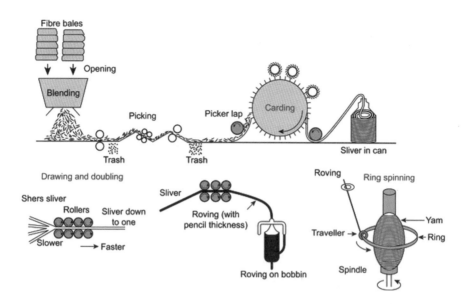

processes in spinning. The main objective of the blow room is to open, mix, clean and even feeding to the next process, which is carding. In the spinning process the cotton arrives in bales(棉包). The fibres are compressed into bales form for easy transportation. To make yarns, fibres of similar length and relatively uniform, fibres from a variety of production lots (组, 批), or fields, or animals must be blended together. The cotton bales arrive in the spinning mill contains a lot of **debris**, dust and leaf particles, which needs to be removed before proceeding to the next stage. At this stage, because the fibres are compressed the **impurities** get enclosed between the compressed fibres, to remove it the fibres need to be opened. As the fibres are being opened the impurities get separated and fall down due to their high weight comparing to the fibres and won't be carried through the ducts by the air.

3.2 Carding

Carding is the next process in the spinning line after the opening and cleaning process. Here the fibres are being straightened and made into the **sliver**. First time in the spinning process the fibres resembles to the thick rope like strand. Carding is an operation where the tufty(丛生的, 簇生的) condition of the fibres is converted into an individual fibre form. The separation of fibres in individual form is one fundamental operation of carding while the other fundamental operation is the formation of the **card sliver**. Carding is a very important process because unless the fibres are separated into individuals, they cannot be spun into smooth and uniform yarns, neither can they be blended properly with other fibres. As the blow room only opens the fibre mass from larger tufts to small ones, the main objective of the carding machine is to further open up the smallest tufts into an individual fibre form. The removal of impurities is also an important objective carried out by the card. Since not all the impurities are removed by the blow room, it is essential at the carding to remove the remaining impurities.

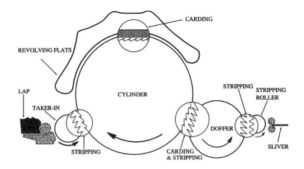

3.3 Drawing

Drawing is the simplest process, a number of slivers from the carding are stretched to make one sliver. This process is purely done to improve the **evenness** of the end product. In order to produce a strong and **uniform** yarn it is necessary to straighten and **align** the fibres and to improve the evenness of the sliver. All of these objectives are achieved by the drawing process carried out by a machine called as the **draw frame**. At the draw frame a number of card slivers are drawn or stretched between several pairs of rollers. As the fibres are attenuated (变细) or drafted, the fibres are straightened and aligned to the axis of the sliver in the direction in which they are drawn.

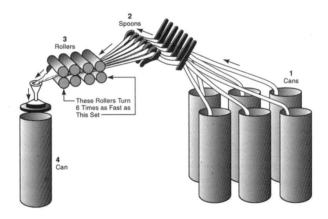

3.4 Combing

Combing is an optional process in the mechanical processing of short staple fibres necessary for the preparation of high quality **combed yarn**. The basic purpose of combing is to remove short fibres and remaining impurities and to make the fibres well aligned and straight so that only high quality long fibres are used for making a yarn. The combing process brings out following positive influence on the yarn characters.

Owing to greater cleanliness and alignment of the fibres, the combed yarns are much smoother and have better lustre as compared to **carded yarns**. Combed yarns are less hairy and compact as compared to carded yarns. Combed yarns can be spun into much finer counts as compared to the carded yarns. The combing process therefore results in an improvement in the quality of the yarn and also enables the spinner to spin finer **yarn counts**. However, the above mentioned quality improvements in the yarn are obtained at a cost of additional machinery, additional process, extra labour and extra floor space. This increases the cost of yarn production considerably.

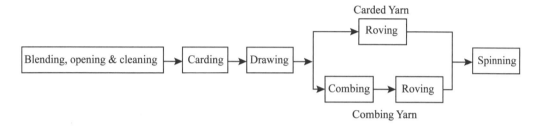

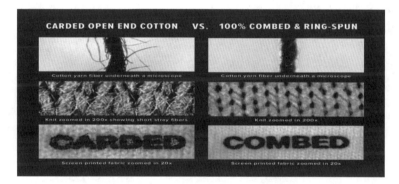

3.5 Roving

This is an intermediate stage before the final yarn is made in the next stage. The problem with the sliver is that it is too thick to put directly into spinning. So an intermediate stage of

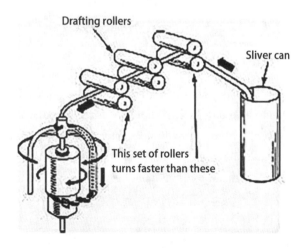

drafting is carried out using the **roving frame**. The draft given at the roving frame reduces the **linear density** of the **drawn sliver** into a less thick strand of fibres suitable as an input to the ring frame. Another advantage of making roving is to have a better package as an input to the ring frame. The roving frame produces roving on compact small packages called as bobbins. The bobbins are much more convenient to transport and have less chances to get damaged as compared to the can sliver mode of package.

3.6 Spinning

The ring spinning is the final process in the formation of the **ring spun yarn**. The basic purpose of the ring spinning frame is to attenuate the roving until the required **fineness** of the yarn is achieved. The roving is taken to the ring frame, which stretches the roving to the required thickness and imparts twists to give strength to the yarn. Stretching is performed by increasing the speed of the roller pairs gradually so the back roller pair runs slower than the middle pair and the middle pair runs slower than the front creating a stretch in the fibre strand. The twists are imparted by the traveller (钢圈) through which the yarn is passed on to the bobbin which rotates at very high speed. The rotation of the bobbin causes the traveller to rotate onto the ring which causes the twists in the fibres leaving the front roller.

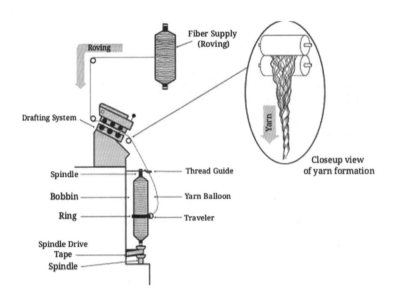

Ring spinning is still the way many fibres are turned into yarn, but in the 20th century another method was invented. Today, some yarns are made through **rotor or open-ended spinning**. In this method, the fibres are fed into a rotor and blown with air into a system that turns them into yarn without using a spindle. Yarns created by rotor spinning tend to be bulkier and heavier than those made by ring spinning.

4. Yarn linear density

Linear density, **yarn count**, yarn number and yarn size are all expressions of the fineness of a yarn. A yarn's or fibre's fineness cannot be expressed in terms of diametre because its diametre is not stable and uniform along its length. Consequently, it is expressed either by measuring the weight of a known length of yarn or by measuring the length of a known weight of yarn. These two basic methods for expression of linear density of any textile yarn are known as: weight based system and length based system.

4.1 Weight based system / Fixed length system

The conditioned weight in grams of 1,000 metres of yarn is called **tex** of a yarn. When the tex of yarn increases, the yarn becomes coarser. The unit of **decitex** (**dtex**) is commonly used for finer yarns, which is one tenth of the tex.

Denier is most commonly used to express the count of a continuous filament yarn of man-made or regenerated fibres. The weight of yarn in grams of 9,000 metres of yarn is called denier of a yarn. Yarn becomes coarser as the denier of a yarn increases.

4.2 Length based system / Fixed weight system

Cotton count (**English count**) is a common unit in this system used in international trade for cotton yarns or cotton type yarns. The number of cotton count is defined as the number of 840 yard lengths of a conditioned yarn that weighs one pound.

A similar unit to cotton count is the **metric count**, which is defined as the number of metres in length of a conditioned yarn that weighs one kilogram. When the count of yarn increases, the yarn becomes finer.

5. Yarn twist

Twist is the measurement of spiral turns given to a yarn in order to hold the constituent fibres or threads together. Twist binds the fibres together and contributes strength to the spun yarn. The number of twist affects yarn and product performance and yarn cost.

In addition to the amount of twist in a yarn, the direction of the twist is also **designated**. There are two types: Z twist and S twist.

When a yarn is twisted in a clockwise direction, the fibres from a **helical** angle are at the yarn surface. This angle conforms to the middle part of the letter "Z" and, therefore, is referred to as "Z" twist (right-hand twist). Open end and ring spun yarn can be produce with "Z" twist. Z-twist is more common for weaving yarns.

When a yarn is twisted in an anticlockwise direction, the fibres from a helical angle are at the yarn surface. This angle conforms to the middle part of the letter "S" and, therefore, is referred to as "S" twist (left-hand twist). Ring spun yarn (not open end yarn) can be produce with "S" twist.

The **twist level** (degree of twist) in a yarn is the number of turns or twist per unit length. The amount of twist varies with fibre length, yarn size, and its intended use.

Chapter 4　Textile and Manufacture

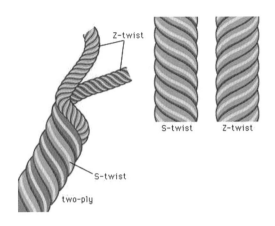

Words:

strand [strænd]	n. (绳子的)股，绞
perpendicularly [ˌpɜːpənˈdɪkjʊləlɪ]	adv. 垂直地，直立地
bobbin [ˈbɒbɪn]	n. 线轴，绕线筒
lubricant [ˈluːbrɪkənt]	n. 润滑剂，润滑油
insulate [ˈɪnsjuleɪt]	vt. 使绝缘，使隔热
bristle [ˈbrɪsl]	n. 鬃毛；刚毛
solution [səˈluːʃn]	n. 溶液；溶解
soluble [ˈsɒljəbl]	adj. [化]可溶的
solvent [ˈsɒlvənt]	n. [化]溶剂，溶媒
volatile [ˈvɒlətaɪl]	adj. (液体或油)易挥发的
non-volatile [ˈnʌnˌvɒlətaɪl]	adj. (液体或油)不易挥发的
filter [ˈfɪltə(r)]	v. 过滤；渗透
condensation [ˌkɒndenˈseɪʃn]	n. (气体)冷凝
sterilize [ˈsterəlaɪz]	vt. 消毒；使无菌
suture [ˈsuːtʃə(r)]	n. 缝合，缝伤口；缝线
dressing [ˈdresɪŋ]	n. 敷料
flock [flɒk]	n. 很短的纤维

velvet ['velvɪt]	n. 丝绒；天鹅绒；立绒
debris ['debriː, 'deɪbriː]	n. 残骸，碎片；垃圾
impurity [ɪm'pjʊərəti]	n. 杂质
carding ['kɑːdɪŋ]	n. 梳理(棉、毛、麻等)；梳毛
combing ['kəʊmɪŋ]	n. 梳毛，梳理
sliver ['slɪvə(r)]	n. 条子，梳条
evenness ['iːvənnəs]	n. 平整度；均匀度
uniform ['juːnɪfɔːm]	adj. 均匀的；一致的
align [ə'laɪn]	v. 对准，校直
fineness ['faɪnnəs]	n. 细度
tex [teks]	n. 特克斯
decitex (dtex) [desi'teks]	n. 分特克斯
denier ['deniə(r)]	n. 旦尼尔
twist [twɪst]	n. 捻 vt. 加捻
designate ['dezɪgneɪt]	v. 命名，指定
helical ['helɪkl]	adj. 螺旋状的

Phrases:

polymer chip	聚合物切片
blow room	吹风室，滤尘室
card sliver	粗梳生条
card frame	梳棉机
draw frame	并条机
roving frame	粗纱机
drawn sliver	熟条
ring spun yarn	环锭纺纱
rotor spinning	转杯纺纱

open-ended spinning	自由端纺纱
yarn count	纱线支数
cotton count (English count)	英制棉纱支数
metric count	公制支数
linear density	线密度
twist level	捻度

Critical reading and thinking

Task 1 Overview

Work in pairs and retell the major processes of ring spinning. Use as many lexical chunks as possible.

Process	Function
opening, cleaning and blending	
carding	
drawing	
combing	
roving	
spinning	

Task 2 Group discussion

Work in groups of 4-5 and have a discussion about the following questions.

1. What are the similarities of and the differences between carded yarn and combed yarn?
2. What are the two mainly methods of spinning synthetic filaments? What are the differences?
3. The thickness or diametre of a yarn is one of its most fundamental properties, but it is impossible to measure it directly. What are the basic methods for expression of linear density of the textile yarn?

Task 3 Language building-up

I. Translate the following terms from English into Chinese or vice versa.

yarn count	
roving frame	
weight based system	
carded yarn	
melt spinning	
捻度	
聚合物切片	
环锭纺纱	
粗梳生条	
短纤纱	

II. Complete the following sentences with the words or phrases given in the previous exercise.

1. In _____, the molten or melted polymer is pushed vertically downwards through the spinneret.
2. _____ consists of short, staple fibres woven together to create a single yarn to be used in weaving, knitting projects, and clothing manufacturing.
3. The _____ expresses the thickness of the yarn, and must be known before calculating the quantity of yarns for a known length of fabric.
4. The _____ (degree of twist) in a yarn is the number of turns or twist per unit length.
5. The separation of fibres in individual form is one fundamental operation of carding while the other fundamental operation is the formation of the _____.

Task 4 Guessing game

Guess the missing lexical chunks from different contexts with the ones in the box.

A. ring spinning	B. combed yarn	C. linear density

Group 1
1. ACP Quality Package therefore becomes an indispensable spinning accessory in the drafting

system of a modern _____ machine.

2. On the traditional 3-roller double-apron drafting system of the cotton _____ frame, the drafting process in the main zone has not been optimal until now.

3. The commonly used, time-tested spinning technique, _____ is one of oldest machine oriented spinning techniques used for staple fibre spinning.

Group 2

1. The _____, a measure of the mass per unit length of a fibre, is used by fibre manufacturers as a measure of fineness.

2. They reported that the bending stiffness and compression of the fabrics produced of yarns with higher _____ are more than the fabrics with lower _____ yarns.

3. Fibre _____ is another important characteristic and has a significant effect on calculating the force acted on the beads which would determine the fibre motion in the jet flow.

Group 3

1. _____ is more costly than carded yarn because the combing process is extremely time-intensive.

2. _____ are highly used for high quality and finer count, used for good quality fabrics.

3. Fabric made of _____ is gentle against the skin.

Task 5 Translation

Translate the following paragraph into Chinese.

In brief, worsted and woolen are different types of wool (long staple vs short staple), prepared in different ways, resulting in a different look and feel. Under magnification, worsted yarns look smooth with long fibres, and woolen yarns are much hairier, with lots of short fibres and more pokey-out bits. Worsted wools are slick when woven, woolen wools are knitted, crocheted, or woven into softer, fluffier fabric, or fulled fabric. Worsted wools are better at keeping out the wind and rain, but woolen wools are warmer, because they are full of air which acts as insulation. Because worsted wool is made from long fibres which all lie parallel, the natural crimp of the wool is removed, and the forms a very tight, hard yarn when spun, with little space between the fibres (as opposed to the more open, fluffy woollen yarn). When woven into fabric, worsted fabric has a tighter, harder, shinier finish, and can make a finer, lighter weight fabric.

Task 6 Research

Surf the Internet for more information about ring spinning and yarn linear density.

(You can try https://www.textileschool.com/349/ring-spinning-process/, https://www.sciencedirect.com/topics/engineering/linear-density.)

Section 3 Weaving and Woven Fabrics

Lead-in

Warm-up questions:

1. Do you know any fabric forming method? What kind of method is used in the fabrics as shown in the following pictures?

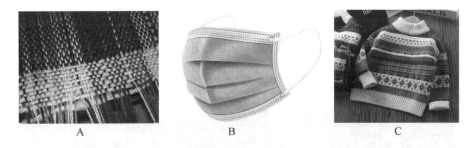

A   B   C

2. What do you know about the traditional and modern weaving loom and weaving technology?

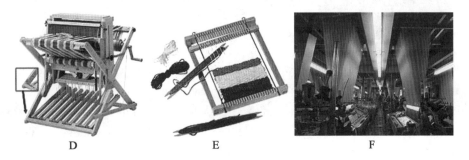

D   E   F

Lexical chunk bank	
引纬	picking/weft insertion
打纬	beating up/in
牵拉	taking up
送经	letting off

119

织机	weaving loom
有梭织机	shuttle loom
无梭织机	shuttleless loom
喷气织机	air jet loom
喷水织机	water jet loom
剑杆织机	rapier loom
片梭织机	projectile loom

There are mainly three kinds of fabric forming methods: the weaving, the knitting and the nonwoven.

The weaving: by **interlacing** two sets of yarns at right angles into fabric.

The knitting: by **intermeshing** the loops of yarns into fabric.

The nonwoven: by bonding fibres, filaments, yarns or combinations of these into fabric.

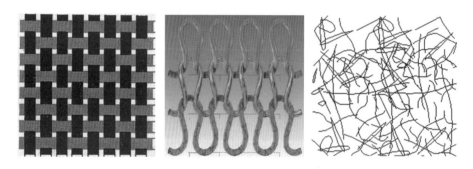

Weaving is the most popular way of fabric manufacturing. It is a method of textile production in which two distinct sets of yarns or threads are interlaced at right angles to form a fabric or cloth. The longitudinal or lengthwise threads are called the warp and the **lateral** or widthwise threads are the weft or filling. Cloth is usually woven on a loom, a device that holds the warp threads in place while filling threads are woven through them.

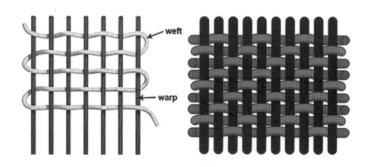

Section 3 Weaving and Woven Fabrics

1. Weaving preparations

1.1 Winding: a process to shift yarn from bobbin/ring into a convenient package more suitable for the next process.

1.2 Warping: a process to transfer the warp yarn from the single yarn package to an even sheet of yarn representing hundreds of ends and then wound onto a warp beam.

1.3 Sizing/Slashing: a process to give the strength to the warp yarn, make it smoother and lubricate it, also reduce the abrasion by giving it through the section of the slasher.

1.4 Drawing-in: a process to draw every warp yarn through its **dropper, heddle eyes** and **reed splits.**

2. Weaving process

Woven fabrics are formed by interlacing the lengthwise warp yarns and widthwise weft yarns. There are a variety of woven structures, but as far as the weaving process is concerned, a woven fabric is formed by the following steps: **shedding, picking, beating up, taking up** and **letting off**.

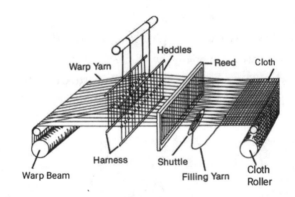

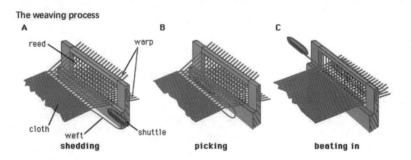

2.1 Shedding: the process of separating the warp yarn into two layers by raising the **harness** to form an open area between two sets of warps and known as shed.

2.2 Picking: the process of inserting the filling yarn through the shed by the means of the shuttle while the shed is opening.

2.3 Beating in/up: the process of pushing the filling yarn into the already woven fabric at a point known as the **fell** and done by the **reed.**

2.4 Taking up: the motion that delivers warp yarn to the weaving area at the required rate and at suitable constant tension by unwinding it from the warp beam.

2.5 Letting off: the motion that withdraws the fabric from the weaving area at the constant rate that will give the required pick-spacing and then winds it onto a roller.

3. Woven fabrics

There are many varieties of woven structures, which can be classified as three types: **primary weaves**, **derivative weaves** and **combination weaves**. The following are the three primary woven structures: **plain weave**, **twill weave** and **satin (or sateen) weave.**

Plain weave is the most common woven structure, with a simple alternate interlacing of warp and filling yarns. Any type of yarn made from any type of fibre can be manufactured into a plain weave fabric. Fabrics in this structure are widely used for cotton and polyester—cotton **sheeting** (bed linen) and **shirting**.

Twill weave is a basic weave, which has a **diagonal** effect on the face, or right side of the fabric. In some twill weave fabrics, the diagonal effect may also be seen clearly on the back side of the fabric. This structure is used for **sturdy** products such as **denim, gabardine, serge** and **khaki**. They are commonly used as the **shell fabrics** for uniforms.

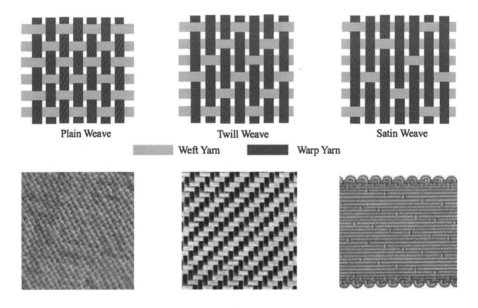

Satin weave is a basic weave, which is characterized by long floats of yarn on the face of the fabric. It can be divided into **warp face satin** and **weft face (filling face) satin**, which is called sateen. Satin weave fabric always has the warp yarns floating over filling yarns. It is commonly used for upholstery, home decorating and fashionable apparel. Satin fabrics are commonly made with silk or man-made filament yarns.

4. Weaving looms

The weaving machine, or loom, has been in use since 4400 BC. For many years, weaving technology and weaving machinery have seen tremendous developments. However, irrespective of whether they are modern or traditional, all looms have the same fundamental features.

Different types of looms are most often defined by the way that the weft, or pick, is inserted into the warp. Many advances in weft insertion have been made in order to make manufactured cloth more cost effective. There are several types of weft insertion and they are as follows:

4.1 Shuttle looms

The shuttle loom is the oldest type of weaving loom which uses a shuttle which contains a bobbin of filling yarn that appears through a hole situated in the side. The shuttle is batted or hit across the loom and during this process, it leaves a trail of the filling at the rate of about 110 to 225 **picks** per minute (ppm). Although very effective and versatile, the shuttle looms are slow and noisy. Also, the shuttle sometimes leads to **abrasion** on the warp yarns and at other times causes thread breaks. As a result, the machine has to be stopped for tying the broken yarns.

4.2 Shuttleless looms

The use of shuttleless looms is important for adding fabrics, adjusting fabric structure, reducing **fabric defects**, improving fabric quality, reducing noise, and improving working conditions. The shuttleless loom has a high speed and is usually 4-8 times more efficient than the shuttle loom.

Many kinds of shuttleless looms are used for weaving such as air jet loom, water jet loom, rapier loom and projectile loom.

4.2.1 Air jet loom

An air-jet loom uses short quick bursts of compressed air to **propel** the weft through the shed in order to complete the weave. Air jets are the fastest traditional method of weaving in modern manufacturing and they are able to achieve up to 1,500 picks per minute. However, the amounts of compressed air required to run these looms, as well as the complexity in the way the air jets are positioned, make them more costly than other looms.

4.2.2 Water jet loom

Water-jet looms use the same principle as air-jet looms, but they take advantage of

pressurized water to propel the weft. The advantage of this type of weaving is that water power is cheaper where water is directly available on site. Picks per minute can reach as high as 1,000.

4.2.3 Rapier loom

This type of weaving is very versatile, in that rapier looms can weave using a large variety of threads. There are several types of rapiers, but they all use a hook system attached to a rod or metal band to pass the pick across the shed. These machines regularly reach 700 picks per minute in normal production.

4.2.4 Projectile loom

Projectile looms utilize an object that is propelled across the shed, usually by **spring power**, and is guided across the width of the cloth by a series of reeds. The projectile is then removed from the weft fibre and it is returned to the opposite side of the machine so it can be reused. Multiple projectiles are in use in order to increase the pick speed. Maximum speeds on these machines can be as high as 1,050 ppm.

Words:

interlace [ˌɪntəˈleɪs]	v. (使)交错，(使)交织
intermesh [ˌɪntəˈmeʃ]	v. (使)互相结合，(使)互相啮合
lateral [ˈlætərəl]	adj. 横向的；侧面的
winding [ˈwaɪndɪŋ]	n. 络筒
warping [ˈwɔːpɪŋ]	n. 整经
sizing [ˈsaɪzɪŋ]	n. 上浆
slashing [ˈslæʃɪŋ]	n. 上浆
drawing-in [dˈrɔːɪŋˈɪn]	n. 穿经
dropper [ˈdrɒpə(r)]	n. 停经片
shedding [ˈʃedɪŋ]	n. 开口
picking [ˈpɪkɪŋ]	n. 引纬
harness [ˈhɑːnɪs]	n. 综框
heddle [ˈhedl]	n. 综丝；综片综线
fell [fel]	n. 织口
reed [riːd]	n. 钢扣
sheeting [ˈʃiːtɪŋ]	n. 粗平布，被单料子

shirting [ˈʃɜːtɪŋ]	n. 细平布，衬衫衣料
diagonal [daɪˈægənl]	adj. 斜纹的；对角线的；斜线的
sturdy [ˈstɜːdi]	adj. 坚固的，耐用的
denim [ˈdenɪm]	n. 斜纹粗棉布，牛仔布
gabardine [ˌgæbəˈdiːn]	n. 一种斜纹防水布料，华达呢
serge [sɜːdʒ]	n. 毛哔叽
khaki [ˈkɑːki]	n. 卡其布
pick [pɪk]	n. 纬纱；引纬
abrasion [əˈbreɪʒn]	n. 摩擦
propel [prəˈpel]	v. 推进；推动
pressurized [ˈpreʃəraɪzd]	adj. 增压的；加压的

Phrases：

heddle eye	综眼
reed split	筘隙
primary weaves	基本机织组织
derivative weaves	变化机织组织
combination weaves	复合机织组织
plain weave	平纹组织
twill weave	斜纹组织
satin weave	经面缎纹
sateen weave	纬面缎纹
shell fabric	面料
warp face satin	经面缎纹
weft face (filling face) satin	纬面缎纹
weaving loom	织机
weaving technology	织造工艺

fabric defect	织疵
spring power	弹簧动力

Critical reading and thinking

Task 1　Overview

Work in pairs and retell the five major motions of weaving process. Use as many lexical chunks as possible.

Weaving process	Function
shedding	
picking	
beating in/up	
taking up	
letting off	

Task 2　Group discussion

Work in groups of 4-5 and have a discussion about the following questions.

1. What are the weaving preparations before weaving? Why are they necessary?
2. What are the three primary woven structures?
3. What are the advantages and disadvantages of the shuttle loom and shuttleless loom?

Task 3　Language building-up

Ⅰ. Translate the following terms from English into Chinese or vice versa.

satin weave	
projectile loom	
shell fabric	
letting off	

warp beam	
有梭织机	
打纬	
平纹组织	
织造工艺	
喷气织机	

II. **Complete the following sentences with the words or phrases given in the previous exercise.**

1. A process to transfer the warp yarn from the single yarn packages to an even sheet of yarn representing hundreds of ends and then wound onto a _____.
2. _____: The process of pushing the filling yarn into the already woven fabric at a point known as the fell and done by the reed.
3. _____ uses short quick bursts of compressed air to propel the weft through the shed in order to complete the weave.
4. _____ fabric always has the warp yarns floating over filling yarns. It is commonly used for upholstery, home decorating and fashionable apparel.
5. A _____ is the outer layer of warm clothing, often the outer layer of an insulated jacket, or it can be a single-layer outer jacket or shirt.

Task 4 Guessing game

Guess the missing lexical chunks from different contexts with the ones in the box.

A. shuttleless loom	B. weft insertion	C. twill weaves

Group 1

1. A _____ is created by passing the weft thread over two or more warp threads and then repeating that pattern one warp thread over, so that a diagonal line is formed.
2. _____ also can be defined as "The order of interlacing which causes diagonal lines of warp and weft floats to be formed on the cloth".
3. The main thing that differentiate _____ from other weaves is the prominent diagonal line which runs along the width of the fabric.

Group 2

1. Rapid warp preparation, a high _____ rate, minimal production disruptions and

flawless fabric quality. With Groz-Beckert healds all these benefits come together for optimum results in day-to-day weaving operation.

2. _____ was originally created for wall lining and curtain applications because of its superior strength, low cost and wide width capability.

3. The shuttle loom is a weaving machine which uses the traditional shuttle (wooden shuttle or plastic shuttle) _____.

Group 3

1. The _____'s basic feature is to separate the weft package from the shuttle or carry only a small amount of weft and replace the large and heavy shuttle with a small and light weft inserter provides favorable conditions for high-speed weft insertion.

2. _____ are of great significance for increasing fabric varieties, adjusting fabric structure, reducing fabric defects, improving fabric quality, reducing noise, improving working conditions, and reducing energy consumption.

3. At present, _____ in developed industrialized countries have replaced shuttle looms.

Task 5 Translation

Translate the following paragraph into Chinese.

Jacquard loom, also called Jacquard Attachment, or Jacquard Mechanism, in weaving, device incorporated in special looms to control individual warp yarns. It enabled looms to produce fabrics having intricate woven patterns such as tapestry, brocade, and damask, and it has also been adapted to the production of patterned knitted fabrics. The Jacquard system was developed in 1804-1805 by Joseph-Marie Jacquard of France, but it soon spread elsewhere. His system improved on the punched-card technology of Jacques de Vaucanson's loom (1745). Jacquard's loom utilized interchangeable punched cards that controlled the weaving of the cloth so that any desired pattern could be obtained automatically. These punched cards were adopted by the noted English inventor Charles Babbage as an input-output medium for his proposed analytical engine and were used by the American statistician Herman Hollerith to feed data to his census machine. They were also used as a means of inputting data into digital computers but were eventually replaced by electronic devices.

Task 6 Research

Surf the Internet for more information about weaving loom and weaving technology.
(You can try https://www.fibre2fashion.com/industry-article/7411/basics-of-weaving-technology-and-modern-looms, https://www.textileschool.com/246/basics-weaving-woven-fabrics/.)

Section 4　Knitting and Knitted Fabrics

Lead-in

Warm-up questions:

1. Do you have any experience of hand knitting?

2. All of the fabrics in the following pictures belong to knitwear. However, they are different knitted fabrics. Can you tell the differences among them?

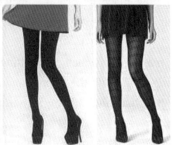

Lexical chunk bank	
纬编针织	weft knitting
经编针织	warp knitting
平针织物	plain fabric
罗纹织物	rib fabric
双罗纹织物	interlock fabric
双反面织物	purl fabric
特利考经编机	Tricot machine
拉舍尔经编机	Raschel machine
编链组织	pillar/chain stitch
经平组织	tricot stitch
经缎组织	atlas stitch

1. The introduction of knitting

Knitting is the process of using two or more needles to **loop** yarn into a series of interconnected loops in order to create a finished garment or some other type of fabric. The word is derived from *knot*, thought to originate from the Dutch verb *knutten*, which is similar to the Old English *cnyttan*, "to knot". Its origins lie in the basic human need for clothing for protection against the **elements**. More recently, hand knitting has become less a necessary skill and more a hobby.

Knitting is a technique of producing fabric from a strand of yarn or wool. Unlike weaving, knitting does not require a loom or other large equipment, making it a valuable technique for nomadic and **non-agrarian** peoples. The oldest knitted artifacts are socks from Egypt, dating from the 11th century CE. They are a kind of very fine **gauge**, done with complex color work and some have a short row heel, which necessitates the purl stitch. These complexities suggest that knitting is even older than the **archeological** record can prove. Most histories of knitting place its origin somewhere in the Middle East, and from there it spread to Europe by Mediterranean trade routes and later to the Americas with European **colonisation**.

The **stocking frame** or mechanical knitting machine was invented in 1589 by William Lee, an English clergyman. With the improvement of steam-powered knitting machines in the mid-nineteenth century, machine knitting increasingly shifted to factories to accommodate the larger machines.

The 1920s saw a vast increase in the popularity of **knitwear** in much of the Western world. Knitwear, especially **sweaters/pullovers** became essential part of the new fashions, rather than mostly practical garments of associated with particular occupations (e.g., fishermen). Knitwear was often associated with sport and leisure. High fashion also embraced knitwear, with **Coco Chanel** making prominent use of it and ***Vogue*** magazine featuring patterns.

Production and consumption of knitted fabrics increased dramatically between 1866 and the mid-1950s. However, the early 1980s saw a return to woven fabrics, particularly for outer wearing apparel.

The 21st century has seen a **resurgence** of knitting. This resurgence coincided with the growth of the Internet, as well as the general "Handmade Revolution" and interest in DIY crafts. There are several reasons for consumer acceptance of knitted fabrics. Knits can be made rapidly, so yarn-to-fabric expenses are much lower and quality fabrics can be produced at comparatively low cost. The increase in travel, especially by air, resulted in the need for lightweight and comfortable clothes that require little care and maintain their neat appearance after sitting or packing. Knitted fabrics fit these needs. The tendency for knits to resist wrinkling has been an important factor in their acceptance and continued use.

Knitting is commonly defined as forming a fabric by means of **interlooping** yarns, and may be divided into two types according to the formation method: **warp knitting** and **weft knitting**.

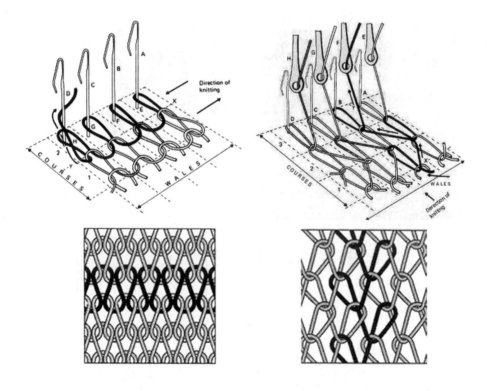

2. Weft knitting and weft knitted fabrics
2.1 Weft knitting

Weft knitting is construction process in which the fabric is made by a yarn forming loops across the width of the fabric or around a circle. **Yarn feeding** and **loop formation** will occur at each needle in succession across the **needle bed** during the same knitting cycle. Works with one yarn at a time run in a horizontal direction. Hand knitting is an example of weft knitting, which can of course be done much faster by machine.

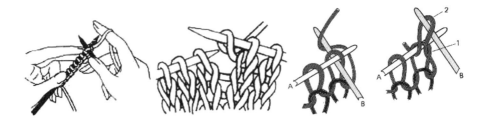

Loop: The simplest unit of knitted structure. It consists of **needle loop** and **sinker loop**. The needle loop include a **head** and two **side limbs/pillars**.

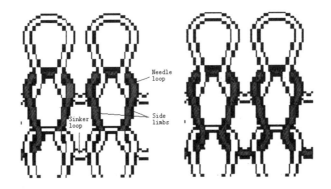

Face loop stitch: The side of the stitch shows the new loop coming through towards the viewer as it passes over and covers the head of the old loop. Face loop tend to show the side limbs of the loops as a series of intermeshing 'V's.

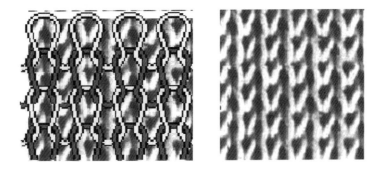

Reverse loop stitch: This is the opposite side of the stitch to the face loop side and shows the sinker loops in weft knitting and the **underlaps** in warp knitting.

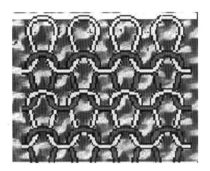 

Single faced fabric: It is produced in warp and weft knitting by the needles operating as a single set. It shows the face loops in one side and the reverse loops in another side.

Double faced fabric: It is produced in warp and weft knitting when two sets of independently controlled needles are employed. It shows the face loops or the reverse loops in both side.

Course: A course is a predominantly horizontal row of loops produced by **adjacent** needles during the same knitting cycle. In weft knitted fabrics a course is composed of yarn from a single supply.

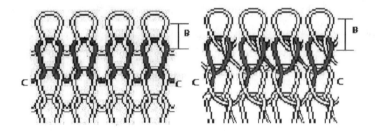

Wale: A wale is a predominantly vertical column of needle loops produced by the same needle knitting at successive knitting cycles and thus intermeshing each new loop through the previous loop. In warp knitting a wale can be produced from the same yarn.

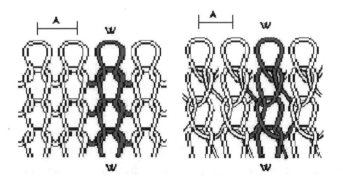

Machine gauge: The "machine gauge" is determined as the number of needles in one inch of needle bed i. e. 2.54 cm. Machine gauge influences choice of yarn and count, and affects fabric properties such as appearance and weight.

Three types of needles: Bearded/Spring Needle, Latch Needle, Compound Needle.

1-stem
2-head
3-beard
4-eye
5-butt
6-tip

1-stem
2-hook
3-latch
4-rivet
5-butt

1-needle
2-tongue
/slider

Knitting action of the Bearded/Spring Needle:

(1) **Clearing:** The old loop is cleared from the hook to the **stem** below the tip of the **beard**.

(2) **Feeding:** A new piece of yarn is fed onto the stem and brought into **hook** by the **sinker wheel**.

(3) **Closing:** The presser presses the beard and the tip of beard enter the **eye cut** in stem. The new yarn therefore is enclosed by beard.

(4) **Landing:** The old loop moves upwards and is located on the outside of the beard as soon as the beard is closed.

(5) **Knocking-over** and loop length formation: As the old loop continues upwards the old loop slide off the needle and the yarn is drawn through it forming a new loop.

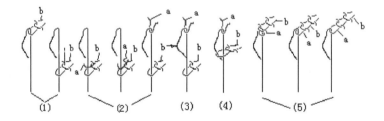

2.2 Weft knitted fabrics

There are four primary base structures of weft knitted fabrics: plain, rib, interlock and purl, from which all weft-knitted fabrics are derived. Each primary structure may exist alone, in a modified form, with stitches other than normal cleared loops, or in combination with another primary structure in a garment length sequence.

Plain fabric (Single jersey)

Plain is produced by the needles knitting as a single set, drawing the loops away from the **technical back** and towards the **technical face** side of the fabric. It is the base structure of ladies' **hosiery**, fully fashioned knitwear and single jersey fabrics. Plain is composed entirely of face loops (or entirely of back loops). Its basic structure unit is only one face loop (or one back loop). The technical face is smooth, with the side limbs of the needle loops having the appearance of columns of Vs in the wales. The technical back has an appearance of columns of semi-circles formed by the heads of the needle loops and the bases of the sinker loops. The edges of the fabric tend to **curl** or **roll**.

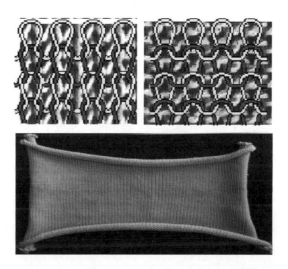

The fabric can be unraveled(脱散), course by course from the course knitted last or from the course knitted first(see Fig. 1). A run/laddering(梯脱, collapse of a wale) will occur if a cut or exposed loop is stressed. The direction of collapse can be either from top to bottom or vice versa(see Fig. 2).

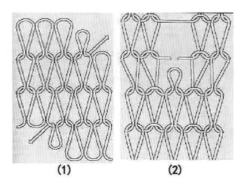

(1)　　　　(2)

Rib fabric

Rib requires two sets of needles operating in between each other so that wales of face stitches and wales of reverse stitches are knitted on each side of the fabric. The simplest rib fabric is 1×1 rib. It consists of alternate face and back wales, where a face wale is composed entirely of face loops and a back wale is composed entirely of back loops. The appearance of the face and back are identical (1×1rib, 2×2rib). The **extensibility** of the fabric widthwise is approximately twice that of single jersey. The lengthwise extensibility is essentially the same as in single jersey.

The fabric does not curl at the edges, therefore it is particularly suitable for the extremities of articles such as tops of socks, the **cuffs** of sleeves, necks, rib borders for garments, and **strapping** for cardigans.

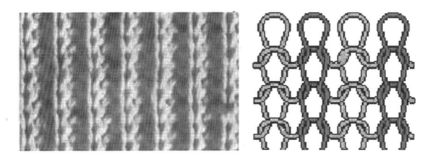

Interlock fabric

Interlock is composed of two 1×1 rib fabrics locked together. It has the technical face of plain fabric on both sides but its smooth surface cannot be stretched out to reveal the reverse loop wales. The appearance of the face and back is the same. Extensibility widthwise and lengthwise are approximately the same as single jersey. The fabric does not curl at edges. The thickness of the fabric is approximately twice that of single jersey, so it is suitable for winter underwear since it has a good **heat retention property**.

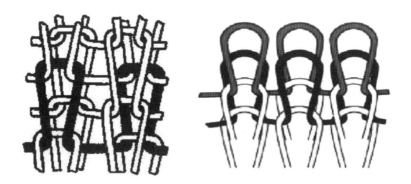

Purl fabric

Purl structures have one or more wales which contain both face and reverse loops. The semi-circles of the needle and sinker loops produced by the reverse loop intermeshing tend to be prominent on both sides of the structure. The appearance of the face and back are identical, very similar to the back of single jersey. The fabric is highly extensible in all directions, approximately twice as extensible as single jersey in the length direction. The fabric tends to be two or three times thicker than single jersey, and does not curl at the edges.

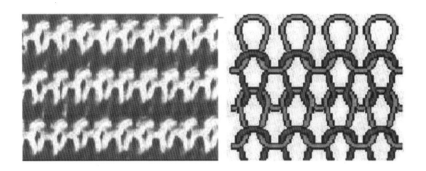

3. Warp knitting and warp knitted fabrics
3.1 Warp knitting

In warp knitting, the work progresses length-wise, through the intermeshing of loops in the direction of wale. In this process, all the loops on the knitting needles in the column remain in the knitting mode. The main advantage of warp knitted cloth is that, unlike weft knitted fabric, it is not easy to unravel. However, these fabrics are not as **stretchable** as the weft knitted fabrics.

Two types of warp-knitting machines are used: Tricot machine and Raschel machine.

Tricot machine

A Tricot machine is a warp knitting machine which uses a single set of bearded or compound needles. The fabric is removed from the needles at approximately 90 degrees. The tricot machine tends to have fine gauge (28-32npi) and fewer **guide bars** (2, 3 or 4), produce simple and fine structure.

Raschel machine

A Raschel machine is a warp knitting machine which uses a single set of vertically mounted latch or compound needles. The fabric is removed from the needles at approximately 150 degrees. The Raschel machine tends to have coarse gauge (16-18npi) and more guide bars (6, 8, 12-48), produce normally coarse and complex structure.

3.2 Warp knitted fabrics

There are three primary base structures of warp knitted fabrics: single bar chain stitches, tricot stitches and atlas stitches.

Pillar/Chain stitches

In the pillar or chain stitch, the same guide always overlaps the same needle. This lapping movement will produce chains of loops in unconnected wales, which must be connected together by the underlaps of a second guide bar. Generally, pillar stitches are made by front guide bars, either to produce vertical stripe effects or to hold the inlays of other guide bars into the structure.

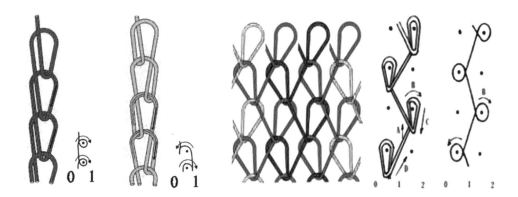

Tricot stitches

Each warp will form loops alternately on two neighbouring needles. Each warp makes at least one needle space underlap after a loop is formed on one needle, and thus the next loop can be formed on an adjacent needle, and all loops can be linked together to form a fabric.

Atlas stitches

The guides carrying warps make atlas lapping. This is a movement where the guide bar **laps** progressively in the same direction for a minimum of two consecutive courses, normally followed by an identical lapping movement in the opposite direction. Usually, the progressive lapping is in the form of **open laps** and the change of direction course is in the form of a **closed lap**, but these roles may be reversed. From the change of direction course, tension tends to cause the heads of the loops to incline in the opposite direction to that of the previous lapping progression. The change of direction course is normally tighter and the return progression courses cause reflected light to produce a faint, **transverse** shadow, stripe effect.

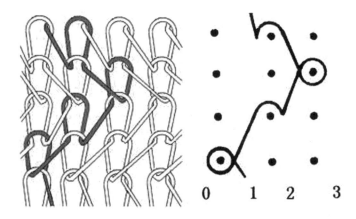

Words：

loop [luːp]	n. 线圈　v. (使)成圈；以环连接
elements [ˈelɪmənts]	n. (尤指恶劣的)天气
non-agrarian [nɒn əˈɡreərɪən]	adj. 非农业的
gauge [ɡeɪdʒ]	n. 机号
necessitate [nəˈsesɪteɪt]	vt. 使……成为必要，需要；强迫，迫使
archeological [ˌɑːkɪəˈlɒdʒɪkəl]	adj. 考古学的
colonisation [ˌkɒlənaɪˈzeɪʃn]	n. 殖民地化
sweater [ˈswetə(r)]	n. 毛衣，运动衫
pullover [ˈpʊləʊvə(r)]	n. 套头毛衣；套头衫
cardigan [ˈkɑːdɪɡən]	n. 开襟绒线衫；羊毛背心；羊毛衫
knitwear [ˈnɪtweə(r)]	n. 针织品
resurgence [rɪˈsɜːdʒəns]	n. 复苏，复活；中断之后的继续
interloop [ˌɪntəˈluːp]	v. 圈套在一起
underlap [ˌʌndəˈlæp]	n. 延展线
course [kɔːs]	n. 横列
wale [weɪl]	n. 纵行
adjacent [əˈdʒeɪsnt]	adj. 相邻的
clearing [ˈklɪərɪŋ]	n. 退圈
feeding [ˈfiːdɪŋ]	n. 垫纱
closing [ˈkləʊzɪŋ]	n. 闭口
landing [ˈlændɪŋ]	n. 套圈
knocking-over [nɒkɪŋˈəʊvə(r)]	n. 脱圈
stem [stem]	n. 针杆
head [hed]	n. 针头
beard [bɪəd]	n. (钩针)针钩
eye [aɪ]	n. 针眼

butt [bʌt]	n. 针踵
tip [tɪp]	n. 针尖
hook [hʊk]	n. 针钩
latch [lætʃ]	n. 针舌；门闩，插销
rivet ['rɪvɪt]	n. 针舌销；铆钉
tongue [tʌŋ]	n. （复合针）针芯
slider ['slaɪdə(r)]	n. （复合针）针芯
presser ['presə]	n. 压板
hosiery ['həʊzɪəri]	n. 袜类；针织内衣
curl [kɜːl] / roll [rəʊl]	v. 卷边
unravel [ʌn'rævl]	n. 脱散
extensibility [ɪksˌtensə'bɪlɪti]	n. 延伸性
cuff [kʌf]	n. 袖口
rib [rɪb]	n. 罗纹
strapping ['stræpɪŋ]	n. 带子
stretchable ['stretʃəbl]	adj. 有弹性的
lap [læp]	n. 垫纱
transverse ['trænzvɜːs]	adj. 横向的；横断的

Phrases：

stocking frame	织袜机
Coco Chanel	可可·香奈儿
Vogue [vəʊg] magazine	《时尚》杂志
yarn feeding	喂纱
loop formation	成圈
needle bed	针床
needle loop	针编弧

sinker loop	沉降弧
side limb/pillar	圈柱
face loop stitch	正面线圈组织
reverse loop stitch	反面线圈组织
single faced fabric	单面织物
double faced fabric	双面织物
bearded/spring needle	钩针/弹簧针
latch needle	舌针
compound needle	复合针
sinker wheel	沉降轮
eye cut	针舌槽
technical back	工艺反面
technical face	工艺正面
heat retention property	保暖性
guide bar	导纱梳栉
open lap	开口式垫纱，开口线圈
closed lap	闭口式垫纱，闭口线圈

Critical reading and thinking

Task 1 Overview

Work in pairs and retell the knitting action of the bearded/spring needle. Use as many lexical chunks as possible.

Knitting action	Properties
clearing	
feeding	
closing	

landing	
knocking-over	
loop length formation	

Task 2 Group discussion

Work in groups of 4-5 and have a discussion about the following questions.

1. What are the four primary base structures of weft knitted fabrics?
2. What are the three primary base structures of warp knitted fabrics?
3. What are the difference between the weft knitted fabrics and warp knitted fabrics?

Task 3 Language building-up

Ⅰ. Translate the following terms from English into Chinese or vice versa.

technical back	
Vogue magazine	
heat retention	
single faced fabric	
Tricot machine	
舌针	
针编弧	
织袜机	
喂纱	
经编针织	

Ⅱ. Complete the following sentences with the words or phrases given in the previous exercise.

1. The thickness of the fabric is approximately twice that of single jersey, so it is suitable for winter underwear since it has a good _____.
2. The _____ or mechanical knitting machine was invented in 1589 by William Lee, an English clergyman.
3. Plain is produced by the needles knitting as a single set, drawing the loops away from the

_____ and towards the technical face side of the fabric.

4. In _____, the work progresses length-wise, through the intermeshing of loops in the direction of wale.

5. _____ and loop formation will occur at each needle in succession across the needle bed during the same knitting cycle.

Task 4 Guessing game

Guess the missing lexical chunks from different contexts with the ones in the box.

| A. reverse loop stitch | B. rib fabric | C. yarn feeding |

Group 1

1. There are various types of _____ systems for circular knitting machines.
2. Positive _____ is a system often fitted on circular knitting machines to positively drive the yarn at a fixed rate relative to the surface speed of the needle cylinder.
3. _____ involves either moving the needles past the stationary yarn feed or moving the yarn past the stationary needle bed.

Group 2

1. _____ create a textured look on any garment.
2. _____ is usually used as lining for matching solid colored fabrics.
3. This 4x2 _____ is super stretchy, with four-way stretch, and full-bodied drape for comfort and ease.

Group 3

1. _____ is the opposite side of the stitch to the face loop-side and shows the new loop meshing away from the viewer as it passes under the head of the old loop.
2. _____ shows the sinker loops in weft knitting and the underlaps in warp knitting most prominently on the surface.
3. The _____ side is the nearest to the head of the needle because the needle draws the new loop downwards through the old loop.

Task 5 Translation

Translate the following paragraph into Chinese.

Knits have also been used as artificial blood vessels in the replacement of damaged blood vessels. Large diametre (greater than 6mm in diametre) artificial blood vessel is generally made of woven or knitted fabric, the former has better stable structure, while the elasticity of the latter is better. The materials used mainly are polyester, PTFE, real silk. At present, the main problems existing in the design and application of large, from material selection to

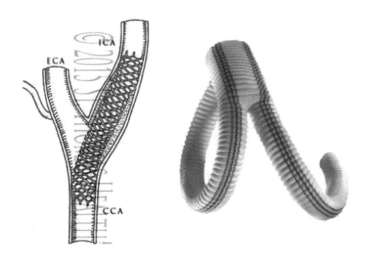

production technology, have been basically solved. Artificial blood vessels with diametre greater than 6 mm have been commercialized while preparation of small caliber vascular is with diametre less than 6 mm is still an international problem. There are two methods to form artificial blood vessel: textile process and nonwoven process. Textile process mainly includes organic weaving and knitting. Among all knitted artificial blood vessel, warp knitting vascular prostheses integrate the advantages of woven and knitted artificial blood vessel, becoming the one most widely used in clinical at present. The nonwoven process mainly has the injection molding and the electrostatic spinning method. No matter is the textile or the nonwoven artificial blood vessel, the properties of antithrombotic, antileakage and biocompatibility have not yet reached the ideal state. The main structural units of the human tissue vessel wall cells are fibrous collagen and elastin, which cause good mechanical properties and adaptability. Generally speaking, artificial vascular materials should have three basic requirements, such as lasting strength, proper pore and good compliance. The basic properties of artificial blood vessels are in order to meet the requirements of anticoagulation and antithrombosis. The research progress is mainly embodied in three aspects: selection of new materials, modification of blood vessels and coating, and artificial vascular endothelium.

Task 6 Research

Surf the Internet for more information about the differences between weft knitting and warp knitting.

(You can try https://textileinsight.blogspot.com/2014/08/weft-knitting.html, https://www.textileschool.com/169/warp-knitting/.)

Section 5 Nonwoven Fabrics

Lead-in

Warm-up questions:

What are the common features of the fabrics below? Are they belonging to woven fabrics or knitted fabrics, or anything else?

Lexical chunk bank	
非织造工艺	nonwoven technology
针刺	needle punching
水刺无纺布	hydroentangled nonwoven fabric
热黏合	thermal bonding
化学黏合	chemical bonding

熔喷工艺	meltblown technology
熔喷无纺布	meltblown nonwoven fabric
纺粘无纺布	spunbonded nonwovens
缝编黏合	stitched bonding

Nonwoven fabrics are broadly defined as sheet or web structures bonded together by **entangling** fibre or filaments (and by **perforating** films) mechanically, **thermally** or chemically. They are flat, porous sheets that are made directly from separate fibres or from molten plastic or plastic film. They are not made by weaving or knitting and do not require converting the fibres to yarn.

Nonwovens are known as **engineered fabrics**, which are manufactured by high-speed and low-cost processes. As compared to the traditional **woven and knitted technology**, a larger volume of materials can be produced at a lower cost by using **nonwoven technology**. The manufacturing principles of nonwovens are manifested in a unique way based on the technologies of creation of textiles, papers, and plastics (Fig. 1), as a result, the structure and properties of nonwovens resemble, to a great extent, to those of three materials.

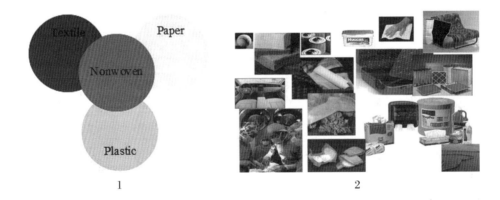

1   2

Nonwovens are already found in many applications, but most are hidden and you do not see them. Fig. 2 displays a few products prepared by using nonwovens.

There are actually many different categories of nonwoven fabrics. According to the production process, non-woven fabrics can be classified as follows.

1. Needle punching process

Needle punching is a nonwoven process by which the fibres are mechanically entangled to produce a nonwoven fabric by repeated penetration of **barbed needles** through a preformed dry fibrous web. The machine which accomplishes this process is known as

needle loom. The picture below displays the **schematic diagram** of a needle loom. The fibrous web, which is unbonded and therefore thick and voluminous, is fed to the machine by a pair of feed rollers.

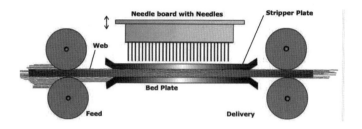

2. Hydroentanglement process

Hydroentangled nonwoven fabric relies on the mechanical entanglement of staple fibres for their coherence. Fibre entanglement is achieved by using jets of water under high pressure instead of barbed needles. The entanglement among the fibres is introduced by the combined effects of the incident water jets and the **turbulent** water created in the web which **intertwines** neighbouring fibres.

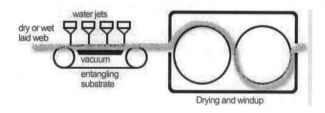

3. Thermal bonding process

It is known that the fibres in the webs can be bonded thermally in order to have sufficient resistance to mechanical deformation. In thermal bonding process, the fibrous web is passed through two **heated rolls or cylinders** pressed against each other. This causes the fibres in the webs to soften and stick together, and, on cooling, bonds are formed within the fibre structure.

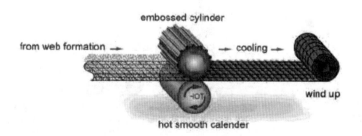

4. Chemical bonding process

In chemical bonding, **chemical binders** (**adhesive materials**) are used to hold the fibres together in a nonwoven fabric. They are applied in a number of different ways to nonwovens and because of their **viscosity** is close to that of water they can easily penetrate into nonwoven structure by **emulsion**. After application of binder by immersion, they are dried and the water evaporates. The binder then forms an **adhesive film** across or between fibre **intersections** and fibre bonding takes place.

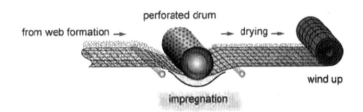

5. Spunbonding process

Spunbonded nonwovens are made by the spunbonding process. When the spinning solution is extruded from the spinneret, **electrostatic charges** and air jets are applied, and the fibres are laid randomly to form a **laminate**. After being heat set by passing the web over a hot cylinder or being reinforced by needle punching, the nonwoven fabric is made.

Spunbond products are employed in carpet backing, **geotextiles**, and **disposable medical/hygiene products**. Since the fabric production is combined with fibre production, the process is generally more economical than when using staple fibre to make nonwoven fabrics.

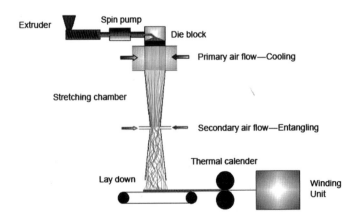

6. Meltblowing process

The **meltblown technology** is based on meltblowing process, where a **thermoplastic** fibre forming polymer is extruded through a linear **die** containing several hundred small **orifices**. **Convergent** streams of hot air rapidly attenuate(衰减) the extruded polymer streams to form extremely fine diametre fibres (1-5 micrometre). The attenuated fibres are subsequently blown by high-velocity air onto a collector conveyor, thus forming a fine fibreed self-bonded **meltblown nonwoven fabric**.

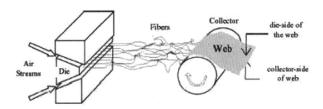

7. Stitched bonding process

Stitched bonding invariably uses a **cross-laid** web, which is fed directly to the **stitch bonder** in a continuous process. The machine used in stitched bonding is basically a modified warp knitting machine which bonds the fabric by knitting columns of stitches down the length of the web.

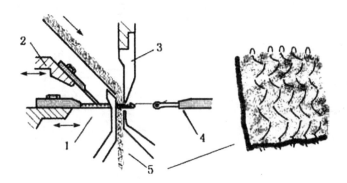

Words:

entangle [ɪnˈtæŋgl]	vt. 使纠缠，缠住
perforate [ˈpɜːfəreɪt]	vt. 穿孔于，在……上打眼
thermal [ˈθɜːml]	adj. 热的，保热的；温热的
hydroentanglement [ˌhaɪdrəɪnˈtæŋglmənt]	n. 水刺(射流缠结)

turbulent [ˈtɜːbjələnt]	adj. 激流的，湍流的
intertwine [ˌɪntəˈtwaɪn]	vt. 缠结在一起；使缠结
adhesive [ədˈhiːsɪv]	n. 黏合剂，黏着剂
viscosity [vɪˈskɒsətɪ]	n. 黏稠；黏性；黏质
emulsion [ɪˈmʌlʃn]	n. 乳状液
intersection [ˌɪntəˈsekʃn]	n. 横断，横切
geotextile [ˈdʒɪəʊtekstaɪl]	n. 土工织物，土工布
thermoplastic [ˌθɜːməʊˈplæstɪk]	n. 热塑性塑料
die [daɪ]	n. 模具
orifice [ˈɒrɪfɪs]	n. 孔；洞口
convergent [kənˈvɜːdʒənt]	adj. 会合的；逐渐减小的
laminate [ˈlæmɪnət]	n. 层压材料；叠层，层压
cross-laid [krɒsleɪd]	adj. 交叉放置的

Phrases：

engineered fabric	工程织物
knitted technology	针织工艺
woven technology	机织工艺
barbed needle	刺针，倒钩针
needle loom	针织(编带)机
schematic diagram	原理图，示意图
heated roll or cylinder	热滚筒
chemical binder	化学黏合剂
adhesive material	黏合材料
adhesive film	胶膜，黏合膜
electrostatic charge	静电荷
disposable medical/hygiene product	一次性医疗/卫生产品
stitch bonder	缝编机

Critical reading and thinking

Task 1 Overview

There are many different categories of non-woven fabrics, work in pairs and list the commonly used methods of nonewoven fabrics. Use as many lexical chunks as possible.

Process	Properties
needle punching process	
hydroentanglement process	
thermal bonding process	
chemical bonding process	
spunbonding process	
meltblowing process	
stitched bonding process	

Task 2 Group discussion

Work in groups of 4-5 and have a discussion about the following questions.
1. What are the properties of nonwoven fabrics?
2. Pleased describe how hydroentangled nonwoven fabrics are made, and how meltblown nonwoven fabrics are constructed.

Task 3 Language building-up

I. Translate the following terms from English into Chinese or vice versa.

meltblown nonwoven fabric	
engineered fabric	
schematic diagram	
chemical binder	

heated roll or cylinder	
非织造工艺	
针刺	
黏合材料	
熔喷工艺	
缝编	

Ⅱ. Complete the following sentences with the words or phrases given in the previous exercise.

1. As compared to the traditional woven and knitted technology, a larger volume of materials can be produced at a lower cost by using _____.
2. _____ invariably uses a cross-laid web, which is fed directly to the stitch bonder in a continuous process.
3. The attenuated fibres are subsequently blown by high-velocity air onto a collector conveyor, thus forming a fine fibreed self-bonded _____.
4. In chemical bonding, _____ (adhesive materials) are used to hold the fibres together in a nonwoven fabric.
5. _____ is a nonwoven process by which the fibres are mechanically entangled to produce a nonwoven fabric by repeated penetration of barbed needles through a preformed dry fibrous web.

Task 4 Guessing game

Guess the missing lexical chunks from different contexts with the ones in the box.

A. thermal bonding   B. hydroentangled nonwoven fabric   C. barbed needles

Group 1

1. The _____ is converted into disposable hospital gowns, surgical curtains and specialty wipes.
2. The main goal of this study is to investigate the performance behavior of _____ in terms of input water jet energy and fabric structural parametres.
3. The high energy of the water removes any dust or short fibres, which explains why _____ have lint-free properties.

Group 2

1. _____, also known as "heat bonding" or melt bonding, is a process that actually melts the web together at fibre crossover points.
2. Thermally bonded carded webs were produced, using these fibres, and characterized in order to understand _____ behavior of fibres with different.
3. The successful bonded microfluidic device was obtained through this optimized _____ method.

Group 3

1. Needle felting is the process of punching _____ through roving laid over a surface fabric.
2. In a needling process, a bed of _____ are repeatedly and rapidly thrust into and pulled out of the layers, thereby entangling fibres of the bats, both with each other and the scrim, and shredding the layer of metallic foil.
3. Two or more layers of fabric are combined by pushing _____ through them.

Task 5　Research

Surf the Internet for more information about nonwoven technology and nonwoven fabrics.

(You can try http://www.technicalnonwovens.com/advantages/technology-advantages, https://www.nonwovens-industry.com/.)

Task 6　Further reading

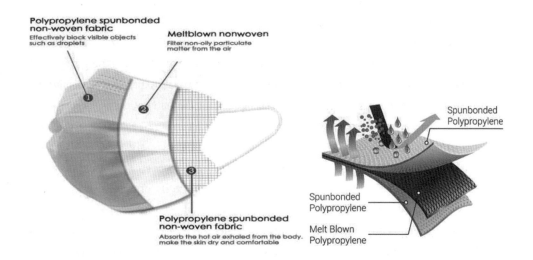

What Is Spunbond Meltblown Fabric and Why Is It in Face Mask?

Spunbond Meltblown Spunbond fabric or SMS has become a hit for the last several months. During the Coronavirus epidemic prevention and control period, masks and disposable clothing are in great demand.

Nonwoven materials have long been used in apparel, health care, engineering, for industrial purposes and for other consumer goods. Spunbond-Meltblown-Spunbond (SMS) is one such material made of Polypropylene. Here we introduce spunbond and meltblown techniques, as well as highlight, the features of SMS and why it is used for face masks.

Spunbond

The spunbond process was first patented in 1940s and with further development has become popular with every decade. The technique itself includes fibres being spun and then directly being dispersed into a web by deflectors or air streams. They use no chemicals, and are thermo-bonded. Spunbonds have low weight, high strength, high air permeability, hydrophilic properties, and excellent wear and tear properties. Now spunbonds are much softer and comfortable than earlier, and weigh between 10 to 150 grams per square metre. They can be easily printed, laminated, dyed, or electrostatically charged.

Meltblown

The meltblown process came after spunbond technology and is the process whereby ultrafine filament (micro-fibres) nonwovens can be produced at low costs. The technique includes hot air being blown onto molten thermoplastic resin that is extruded through a linear die containing hundreds of small holes, to form a fine fibreed self-bonded nonwoven web. Its key feature is that it is an extremely thin fibre. As a result this material is often used for filters for air, liquids and particles, or as absorbents in products such as wipes, oil absorbents, incontinence products, and female hygiene, but can also be used in the production of certain electronics, adhesives, and other apparel.

What is lamination?

Nonwovens lamination is the process of bonding two or more layers, at least one of which is a nonwoven fabric, with the objective of obtaining improved strength, stability, sound insulation, appearance, or other properties. Although less expensive to manufacture, nonwoven fabrics tend to be weaker than woven ones, and the lamination process comes into play just when there is the need to strengthen this nonwoven material for a variety of uses.

Nonwoven lamination creates breathable material that allows moisture to easily pass out of the garment, while preventing liquids from entering. The nonwoven composite material can also have an anti-microbial surface treatment to further enhance its protective properties. The outer layer of nonwoven fabric is soft to the touch, strong and durable.

The bonding is typically done with the use of adhesives (or heat) and pressure, making it possible to manufacture nonwovens products, equipped with a waterproof layer, that are breathable, soft, comfortable and suitable for printing.

Breathability was an issue with lamination in the early days, but the processes used today are capable of producing breathable, porous materials that maintain their waterproof properties. When these are layered to form the SMS material, the features can be combined together and the range of applications increases. Additionally, the combination of spunbond and meltblown materials means that the features of each can make up for the weaknesses of the other. For example, meltblowns have limited strength so can be combined with a spunbond to become a strong material, and likewise spunbonds can be elevated with the addition of meltblowns. These can even be combined to make a material which has a textile feeling to it. SMS has excellent physical properties as well as barrier qualities. Features include high tensile strength, softness, comfort, breathability, wearability, and is also lightweight. It also acts as a water-repellent, and a barrier against bacteria, blood and other liquids as well as gas/steam perspiration. Finally, it is also fine enough to serve as a disposable fabric.

Medical SMS Fabric is therefore suitable for medical and hygiene products such as diapers, protective wear, face masks, hospital gowns, wound care, caps, filtration fabrics, and much more.

Although SMS is more expensive than other materials such as non-layered materials, its unique features and specific design make it widely applicable for use. It can not only mimic the appearance, texture and strength of a woven fabric and can be as bulky as the thickest padding, but in combination with other materials also provides a spectrum of products with diverse properties. With the development of SMS it is thought that it will grow rapidly in the market of nonwovens and become more and more popular for various segments including apparel, geotextiles, furnishings, roofing as well as many others.

Most of hospital gowns, medical caps, medical masks, and shoe covers are manufactured using spunbond and meltblown techniques.

Chapter 5 Textile and Economy

Section 1 Textile and Trade

—The Silk Road

Lead-in

Warm-up questions:

1. What is the Silk Road?
2. How did the Silk Road get its name?
3. How much do you know about the Silk Road?

The **Silk Road** was an ancient network of trade routes connecting China and the Far East with the Middle East and Europe in commerce, formally established during the **Han Dynasty of China** between 130 **BCE**-1453 CE. As the Silk Road was not a single **thoroughfare** from East to West, the term "Silk Routes" has become increasingly favored by historians, though "Silk Road" is the more common and recognized name.

The European explorer **Marco Polo** (1254-1324 CE) traveled on these routes and described them in depth in his famous work but he is not **credited** with naming them. Both terms for this network of roads were **coined** by the German geographer and traveler, **Ferdinand von Richthofen**, in 1877 CE, who designated them *Seidenstrasse* (Silk Road) or *Seidenstrassen* (Silk Route).

Even though the name "Silk Road" derives from the popularity of Chinese silk among tradesmen in the Roman Empire and elsewhere in Europe, the material was not the only important export from the East to the West. Goods transported along the Silk Road economic belt from the West to the East include: horses; saddles and riding tack; the grapevine and grapes; dogs and other animals both **exotic** and domestic; animal furs and skins; honey; fruits; glassware; woolen blankets, rugs, carpets; textiles (such as curtains); gold and silver; camels; slaves; weapons and armor. Goods transported along the Silk Road economic belt from the East to the West include: silk; tea; dyes; precious stones; china (plates, bowls, cups, vases); porcelain; spices (such as **cinnamon** and **ginger**); bronze and gold artifacts; medicine; perfumes; ivory; rice; paper; gunpowder.

While many different kinds of merchandise traveled along the Silk Road, the name comes from the popularity of Chinese silk with the West, especially with Rome. The Silk Road routes stretched from China through India, **Asia Minor**, up throughout **Mesopotamia**, to Egypt, the African continent, Greece, Rome, and Britain.

China and the West

According to the Greek historian Strabo (63BCE-24 CE) the Greeks "extended their empire as far as the Seres". "Seres" was the name by which the Greeks and Romans knew China, meaning "the land where silk came from". It is thought, then, that the first contact between China and the West came around the year 200 BCE.

The Han Dynasty of China (202 BCE-220 CE) was regularly **harassed** by the nomadic tribes of the Xiongnu on their northern and western borders. In 138 BCE, Emperor Wu sent his **emissary** Zhang Qian to the West to negotiate with the **Yuezhi** people for help in defeating the Xiongnu.

Zhang Qian's **expedition** led him into contact with many different cultures and civilizations in central Asia. The consequences of Zhang Qian's journey was not only further contact between China and the West but an organized and efficient horse breeding program throughout the land in order to equip a **cavalry**. With the western horse of the Dayuan, the Han Dynasty defeated the Xiongnu. This success inspired Emperor Wu to **speculate** on what else might be gained through trade with the West and the Silk Road was opened in 130 BCE.

The **Parthians** then became the central **intermediaries** between China and the West.

Eastward Exploration

The Silk Road routes also opened up means of passage for explorers seeking to better understand the culture and geography of the Far East.

Venetian explorer Marco Polo famously used the Silk Road to travel from Italy to China, which was then under the control of the **Mongolian Empire**, where they arrived in 1275.

Notably, they did not travel by boat, but rather by camel following overland routes. They arrived at Xanadu, the lavish summer palace of the Mongolian emperor **Kublai Khan**.

In all, the explorer spent 24 years in Asia, working in Kublai Khan's court. Marco Polo returned to Venice, again via the Silk Road, in 1295, just as the Mongolian Empire was in decline. His journeys across the Silk Road became the basis for his book, ***The Travels of Marco Polo***, which gave Europeans a better understanding of Asian commerce and culture.

The Silk Road Legacy

The network was used regularly from 130 BCE, when the Han officially opened trade with the West, to 1453 CE, when the **Ottoman Empire boycotted** trade with the West and closed the routes.

Camel with Guide, Tang Dynasty by Jan van der Crabben

The Silk Road—from its opening to its closure—had so great an impact on the development of world civilization that it is difficult to imagine the modern world without it. The greatest value of the Silk Road was the exchange of culture. Art, religion, philosophy, language, technology, science, architecture, and every other element of civilization were exchanged along this route, carried with the commercial goods the merchants traded from country to country.

Words:

thoroughfare [ˈθʌrəfeə(r)]	n. 大道，通路
credit [ˈkredɪt]	n. 信用，赞许，学分　vt. (~someone with sth.) 把（某成果）归因于某人
coin [kɔɪn]	vt. 创造（新词语）
exotic [ɪgˈzɒtɪk]	adj. 奇异的
ginger [ˈdʒɪndʒə]	n. 姜
cinnamon [ˈsɪnəmən]	n. 肉桂
harass [ˈhærəs]	vt. 骚扰，侵扰，不断攻击
emissary [ˈemɪsəri]	n. 使者；密使
expedition [ˌekspəˈdɪʃn]	n. 远征；探险队
cavalry [ˈkævlri]	n. 骑兵
speculate [ˈspekjuleɪt]	vi. 推测；投机；思索　vt. 推断
intermediary [ˌɪntəˈmiːdiəri]	adj. 中间的；媒介的；中途的 n. 中间人；仲裁者；调解者；媒介物
boycott [ˈbɔɪkɒt]	vt. 联合抵制；拒绝购买；拒绝参加 n. 联合抵制

Proper Names:

Silk Road	丝绸之路
Han Dynasty of China	汉朝
Marco Polo	马可·波罗
Ferdinand von Richthofen	斐迪南·冯·李希霍芬
Mesopotamia	美索不达米亚

Yuezhi	大宛
Parthian	帕提亚帝国(阿萨息斯王朝或安息帝国)
Mongolian Empire	元朝(蒙古帝国)
Kublai Khan	元世祖(忽必烈汗)
The Travels of Marco Polo	《马可·波罗游记》
Ottoman Empire	奥斯曼帝国
Asia Minor	小亚细亚
BCE: Before the Christ Era	公元前

Critical reading and thinking

Task 1 Overview

Work in pairs and retell the story of Silk Road. Use as many lexical chunks as possible.

Task 2 Group discussion

Work in groups of 4-5 and have a discussion about the following questions.

1. What role did the Silk Road play in China and the West?
2. How did the Silk Road change the world?
3. What is the significance of the "Belt and Road" Initiative put forward by President Xi Jinping?

Task 3 Decide whether the following statements are True or False

1. The Silk Road is the main trade way between East and West in ancient time.
2. The Silk Road stretches from Asia to Europe, African.
3. Among all the goods along the Silk Road, silk influences the Western culture the most.
4. Emperor Wu opened the Silk Road because he wanted to buy horses from the West.
5. The Silk Road existed for about 1,100 years.

Task 4 Translation

Translate the following paragraph into English.

丝绸之路是西汉时期汉武帝派遣张骞出使西域打通的连通东西方的一条主要贸易通道。丝绸之路加强了中国和中亚、欧洲的交流，东西方在物质上互通有无，商业往来频繁。同时还促进了东西方之间的文化交流，深刻地影响了世界文明进程。2013 年中国国家主席习近平以古丝绸之路为蓝本提出了"一带一路"的倡议，迅速得到了全世界 100 多个国家的欢迎和响应。"一带一路"倡议促进了中欧之间政策沟通、设施联通、贸易畅通、资金融通、民心相通，使中欧之间经济、文化交流更加密切。

Task 5　Research

Surf the Internet for more information about the Silk Road.
(You can try https：//www.history.com/topics/ancient-middle-east/silk-road，https：//www.travelchinaguide.com/silk-road/.)

Section 2 The Textile Industry and Economic Recovery

Lead-in

Warm-up questions:

1. How does respiratory epidemic affect economy?
2. What are the roles of textile industry in developing countries or developed countries?

Lexical chunk bank	
新冠肺炎	COVID-19
可支配收入	disposable income
高附加值	high added-value
成本效益	cost-effective
人工智能	artificial intelligence
区块链技术	blockchain technology
劳动密集型行业	labour-intensive sector
零售市场	retail market
资源密集型	resource-intensive
购买力	purchasing power
智能纺织	smart textiles
交货期	lead time
多边开发银行	multilateral development banks

The textiles and apparel sector can create jobs and spur further industrialization in countries recovering from **COVID-19.**

To make the most of this opportunity, however, countries will need to embrace new partnerships and approaches.

The COVID-19 **pandemic** is an unprecedented public health crisis that has exerted an external shock on the global economy. Despite this contraction, the textiles and apparel industry could be a key engine for growth and employment in certain countries.

The textile and apparel opportunity

Countries building back after COVID-19 should not ignore the textiles and apparel industry. It is considered a starter sector in the road to industrialization. When the industry expands, it provides a base on which to build capital for more technologically demanding industries. In fact, the textiles and apparel sector can be critical to the growth and development strategies of many developing countries.

The global textiles and apparel industry market had a **retail** market value of $1.9 trillion in 2019 and is projected by Boston Consulting Group to reach $3.3 trillion in 2030, growing at a compound annual growth rate of 3.5%. Projections ahead of COVID-19 forecast that population growth, rising levels of **disposable income** and rapid urbanization in developing countries would likely drive demand in the future.

Textiles and apparel exports **constitute** an important share of the total exports in a range of countries: 85% in Bangladesh, 59% in Pakistan, 12% in Turkey, and 11% in Egypt. Still, while many countries are well-positioned in the raw materials or the production stage of the textiles and apparel global value chain (GVC), they are only playing a limited role in the absence of retail (comprised of marketing, branding and sales). Thus, they have potential waiting to be unlocked to reap more benefits from the global markets.

Additionally, the textile industry creates a special opportunity during the pandemic given the number of employment opportunities it can provide. This labour-intensive sector employs millions, and the share of employment in the sector across the total manufacturing workforce is significant in Islamic Development Bank (IsDB) member countries.

IsDB's 57 member countries would especially benefit as these countries jointly represent the purchasing power of almost one quarter of the world's population. With GDP growth rates of up to 8% per year, their economies have considerable potential to further increase their market share in the global economy.

Strategic investments

Those investing in the textiles and apparel market could yield **tangible** economic benefits from targeted investments. Sustainable and recycled fibres represent one such opportunity, as they are **poised** to replace resource-intensive raw materials at an increasing pace. Additionally, technical, smart textiles have enormous potential to be used in several industries such as automotive, construction and medical equipment.

Emerging tech, already **transformative**, will continue to shape the textiles and apparel sector. Data applications, artificial intelligence (AI) and machine learning, and 3D printing are some of the technologies enhancing product design processes and reducing lead times, leading to laser-cutting machines, sewing robots, and nanotechnology. Furthermore, COVID-19 has shown the need for blockchain technology to create transparency and **traceability** throughout the supply chain, providing other opportunities within the market.

COVID-19 could also fuel a shift to nearshoring, ensuring factories are closer to their final sales markets. Additionally, leading firms will seek strategic partnerships with first-tier suppliers to meet demand and reduce lead times. The future market structure will mainly be determined by a country's location, as well as the ability of its textiles and apparel industry to provide cost-effective production, competitive skills, quality products and efficient lead times.

Looking ahead

Countries must act swiftly and strategically to restart and restructure their economies. Not doing so could widen existing gaps in wealth, technology, and productivity gaps across borders.

Making these shifts will take a series of tactical changes, including strengthening and broadening processing capabilities, bridging the infrastructure gap, developing sustainable textiles and apparel production capabilities, and strengthening the external image of the country as a destination of choice for the textiles and apparel industry.

Such changes will also take new collaborations between a range of different actors. For textiles and apparel, collaboration with industry associations (within and across countries), as well as joint projects with universities, can strengthen knowledge exchange and drive innovation. Partnerships with brands and knitting houses or weaving mills can also **foster** more vertical integration for companies. These collaborations can unlock higher added-value within countries and provide a strong return on investment.

To this end, the IsDB has set clear goals to **catalyze** private and public investment for its 57 member countries to **fuel** their economic and social development and drive the competitiveness of key industries linked to the global market. For instance, the IsDB is building a platform for its member countries to bring together a range of stakeholders to drive key changes, including: increased trade and investment relations; co-financing arrangements with the public and private sectors; multilateral development banks and other international organizations; and boosted private-sector engagement in development interventions.

Overcoming the hurdles brought by the pandemic—and recovering quickly—will not be easy. It will take new **collaborations** and new approaches that add value to economies. Still, countries and firms that can **adapt** to new shifts and new relationships will reap the most benefits in the medium and long term—and be the most **resilient** ahead of the next big crisis.

Section 2 The Textile Industry and Economic Recovery

Words:

pandemic [pæn'demɪk]	n. 流行病
retail ['riːteɪl]	n. 零售 vi. (以某种价格)零售
constitute ['kɒnstɪtjuːt]	v. 构成
tangible ['tændʒəbl]	adj. 清晰明确的
poise [pɔɪz]	n. 镇静 v. 保持(某种姿势)；抓紧；使稳定
transformative [træns'fɔːmətɪv]	adj. 革命性的；转折的
traceability [ˌtreɪsə'bɪləti]	n. 可追溯性
fuel ['fjuːəl]	vt. 给……提供燃料；刺激，煽动；推动
catalyze ['kætəlaɪz]	vt. 催化；刺激，促进
foster ['fɒstə(r)]	vt. 促进；助长；培养；鼓励
collaboration [kəˌlæbə'reɪʃn]	n. 协作，合作
adapt [ə'dæpt]	vt. 使适应；改编 vi. 适应
resilient [rɪ'zɪliənt]	adj. 弹回的，有弹力的；能复原的；可迅速恢复的

Critical reading and thinking

Task 1 Overview

Work in pairs and retell how textile industry can play role in economic recovery. Use as many lexical chunks as possible.

Task 2 Group discussion

Work in groups of 4-5 and have a discussion about the following questions.

1. What significance does textile industry mean to a country's economy?
2. In the global time, what opportunity can textile trade bring?
3. Why can the epidemic be an opportunity for textile industry?

Task 3 Language building-up

Translate the following terms from English into Chinese or vice versa.

Chapter 5 Textile and Economy

blockchain technology	
compound annual growth rate	
multilateral development banks	
cost-effective	
smart textiles	
零售市场	
购买力	
可支配收入	
资源密集型	
供应链	

Task 4 Research

Surf the Internet for more information about the relation between textile and economy.

(You can try https：//www.weforum.org/agenda/2020/08/, https：//xueshu.baidu.com/usercenter/paper/.)

Section 3　Recyclable Textiles for Green Economy

Lead-in

Warm-up questions:

1. What is green economy?
2. What can textile industry do to boost green economy?
3. What are the benefits of recyclable textiles?

Lexical chunk bank	
不可再生资源	non-renewable resources
取舍模式	take-make-dispose model
全行业的	industry-wide
价值链	value chain
逐步淘汰；分阶段撤销	phase-out
一次性的	one-off
再生农业	regenerative agriculture
化石燃料	fossil fuel

　　It is hard to imagine living in a world without textiles. Nearly everyone, everywhere comes into contact with them nearly all the time. This is especially true of clothing. Clothes provide comfort and protection, and for many represent an important expression of individuality. The textiles industry is also a significant sector in the global economy, providing employment for hundreds of millions around the world. These benefits **notwithstanding**, the way we design, produce, and use clothes has drawbacks that are becoming increasingly clear. The textiles system operates in an almost completely **linear** way: large amounts of **non-renewable resources** are **extracted** to produce clothes that are often used for only a short time, after which the materials are mostly sent to **landfill** or **incinerated**. More than USD 500 billion of value is lost every year due to clothing underutilization and the lack of recycling. Furthermore, this **take-make-dispose model** has numerous negative environmental and societal impacts. For instance, total greenhouse gas emissions from textiles production, at 1.2 billion tons annually, are more than those of all international flights and maritime shipping combined. **Hazardous** substances affect the health of both textile workers and wearers of clothes, and they escape into the environment. When washed, some garments release plastic **microfibres**, of which around half a million tons every year contribute to ocean pollution—16 times more than plastic **microbeads** from cosmetics. Trends point to these negative impacts rising **inexorably**, with the potential for **catastrophic** outcomes in future. This linear system is ripe for **disruption**.

In recent years, the industry and its customers have become increasingly aware of the negative environmental and societal impacts of the current system. Brands and retailers have started to address specific environmental or societal challenges within their supply chains, both individually and through **industry-wide** organizations and initiatives. However, most of these efforts are focused on reducing the impact of the current linear system—for example, by using more efficient production techniques or reducing the impact of materials—rather than taking an upstream, systemic approach to tackling the root cause of the system's wasteful nature directly, in particular, low clothing utilization and low rates of recycling after use.

This report proposes a vision for a new textile economy aligned with the principles of a circular economy: one that is **restorative** and regenerative by design and provides benefits for business, society, and the environment. This vision is distinct from, and complements, ongoing efforts to make the textiles system more sustainable by minimizing its negative impacts.

In such a new textile economy, clothes, fabric, and fibres are kept at their highest value during use, and re-enter the economy after use, never ending up as waste. This would provide a growing world population with access to high-quality, affordable, and individualized clothing, while regenerating natural capital, designing out pollution, and using renewable resources and energy. Such a system would be distributive by design, meaning value is circulated among enterprises of all sizes in the industry so that all parts of the **value chain** can pay workers well and provide them with good working conditions. A new textile economy relies on four **ambitions** (see Fig.1). It would lead to better economic, environmental, and societal outcomes, capturing opportunities missed by the current, linear, textiles system. When implementing these ambitions, each will come with a variety of different solutions for different applications, and their interactions need to be taken into account.

Realizing these ambitions will not happen overnight. While there are some immediate profit opportunities for individual businesses, collaborative efforts across the value chain, involving private and public sector actors, are required to truly transform the way clothes are designed, produced, sold, used, collected, and reprocessed. However, this should not discourage or delay action. The time to act is now, and the ambitions below offer an attractive target state for the industry to align on and innovate towards.

Section 3   Recyclable Textiles for Green Economy

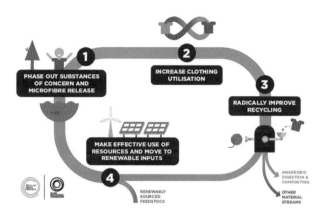

Figure 1   Ambitions for a New Textiles Economy

1. Phase out substances of concern and microfibre release

First and foremost, a new textile economy needs to ensure that the material input is safe and healthy to allow cycling and to avoid negative impacts during the production, use, and after-use phases. This means that substances that are of concern to health or the environment are designed out and no pollutants such as plastic microfibres are **inadvertently** released into the environment and ocean. The following two areas of action could **kick-start** this transition:

—**Align industry efforts and coordinate innovation to create safe material cycles.**

Elimination of substances of concern is needed to enable large-scale recycling, as well as to avoid various negative impacts at all stages of the value chain. Improved transparency along the value chain, a **robust** evidence base, and common standards would facilitate the **phase-out** of substances of concern. While some hazardous substances could be phased out quickly, innovation will be required to create new process inputs (e. g. dyes and **additives**), production processes, as well as textile materials, to fully phase out negative impacts related to substances of concern.

—**Drastically reduce plastic microfibre release.**

New materials and production processes that radically reduce the number of plastic microfibres shed by clothing, alongside technologies that work effectively at scale to capture those that do still shed, are essential for this to be feasible. A better understanding of the causes of microfibre shedding will continue to inform solutions and identify gaps.

2. Transform the way clothes are designed, sold, and used to break free from their

increasingly disposable nature

Increasing the average number of times clothes are worn is the most direct lever to capture value and design out waste and pollution in the textiles system. Designing and producing clothes of higher quality and providing access to them via new business models would help shift the perception of clothing from being a disposable item to being a durable product. As the acts of buying and wearing clothes fulfil a complex array of customer needs and desires, a variety of sales and service models is needed in a new textile economy. Economic opportunities already exist in various market segments, and brands and retailers could exploit these through refocused marketing. The take-up of new opportunities would benefit from collaborative action to stimulate the development of innovative business models. Such action would also help unlock potential where the immediate economic case is not yet evident at scale.

Three areas of action would speed the transition towards this ambition:

Scale up short-term clothing rental.

When garments can be worn more often than a customer is able or willing to do, rental models could provide an appealing business opportunity. For customers desiring frequent outfit changes, subscription-based models can offer an attractive alternative to frequently buying new clothes. For garments where practical needs change over time, for example, children's clothes or those for **one-off** occasions, rental services would increase utilisation by keeping garments in frequent use rather than in people's closets. For all these models, refocused marketing—using the vast experience and capacity that brands and retailers have-and optimised **logistics** are key enablers for stimulating growth of new service offerings.

Make durability more attractive.

While short-term clothing rental can capture the value of durability by distributing clothing use between many different people, for certain clothing types and customer segments, quality and durability can be of value even if there is only one or a few users. In these segments, many customers value high-quality, durable clothes, but a lack of information prevents the full value capture. For clothes that have already been used and become unwanted, but which are still durable enough to be used again, enhanced resale models offer an attractive opportunity. A focus on delivering quality purchases that last longer also encourages new technologies to be exploited that offer better fit and customisation for maximum customer satisfaction.

Increase clothing utilisation further through brand commitments and policy.

Driving high usage rates requires a commitment to design garments that last—an industry transition which could be advanced through common guidelines, aligned efforts, and increased transparency. Policymakers can also have an important role in further increasing clothing utilisation.

3. Radically improve recycling by transforming clothing design, collection, and reprocessing

There is a compelling case for radically improving recycling to allow the industry to capture the value of the materials in clothes that can no longer be used. Increasing recycling represents an opportunity for the industry to capture some of the value in more than USD 100 billion worth of materials lost from the system every year, as well as to reduce the negative impacts of their disposal.

A combination of demand and supply-side measures in the following four areas would be needed to realise this ambition:

—**Align clothing design and recycling processes.** Currently, clothing design and production typically do not consider what will happen when clothes cannot be used anymore. **Converging** towards a range of materials (including blends where those are needed for **functionality**), and developing efficient recycling processes for these, is a crucial step in scaling up recycling, as is the development of new materials, where current ones do not provide the desired functionality and recyclability. Alignment is also needed to provide tracking and tracing technologies to identify materials in the recycling process.

—**Pursue technological innovation to improve the economics and quality of recycling.** Existing recycling technologies for common materials need to drastically improve their economics and output quality to capture the full value of the materials in recovered clothing. A shared innovation agenda is needed to focus efforts and investments towards recycling technologies for common materials. Improved **sorting** technologies would also support increased quality of recycling by providing well-defined **feedstock**, in particular in the transition phase until common tracking and tracing technologies exist.

—**Stimulate demand for recycled materials.**

Increasing demand for recycled materials through clear commitments to using more recycled input could drastically accelerate the uptake of clothing recycling. Better matching supply and demand through increased transparency and communication channels, as well as policy, would further help stimulate demand.

—**Implement clothing collection at scale.**

Clothing collection needs to be scaled up dramatically alongside recycling technologies and, importantly, implemented in locations where it currently does not exist. Creating demand for recycled materials will increase markets for non-wearable items, dramatically improving the opportunity for collectors to capture value from these materials. Guidelines on comprehensive collection—based on current best practices and further research on optimal collection systems—would help scale up collection. These guidelines should include a set of global collection archetypes, allowing for regional variation but building on a set of common principles.

4. Make effective use of resources and move to renewable inputs

The need for raw material inputs in a new textile economy would be drastically reduced due to higher clothing utilisation and increased recycling. However, virgin material input will likely always be required. Where such input is needed and no recycled materials are available, it should increasingly come from renewable resources. This means using renewable feedstock for plastic-based fibres and **regenerative agriculture** to produce any renewable resources.

In addition, transitioning to more effective and efficient production processes—that generate less waste (such as **offcuts**), need fewer inputs of resources, such as **fossil fuels** and chemicals, reduce water use in water-scarce regions, are energy efficient, and run on renewable energy—can further contribute to reducing the need for non-renewable resource input. Accounting for and reporting the costs of negative **externalities** would further support the shift to better resource use and production processes, and thereby generate system-wide benefits.

The new textile economy would have the following characteristics:

1. **A new textile economy produces and provides access to high-quality, affordable, individualized clothing.**

   In a new textile economy, everyone has access to the clothes that they need, when they need them. New business models allow more flexibility on which clothes to wear and when, as well as provide access to clothes that might not be affordable through traditional sales. Clothes are designed and produced to provide high quality, durability, and flexibility—for example, in the form of individualized or modifiable clothes.

2. **A new textile economy captures the full value of clothing during and after use.**

   In a new textile economy clothes are used more often, allowing their value to be captured fully. Once clothes are not used anymore, recycling them into new clothes allows the value of the materials to be captured at different levels. For this to be successful, no substances of concern that could contaminate products and prevent them from being safely recycled are used.

3. **A new textile economy runs on renewable energy and uses renewable resources where resource input is needed.**

   The energy required to fuel a new textile economy is renewable by nature, decreasing resource dependence and increasing system resilience. Resources are kept in the system and where input is needed, this comes from renewable resources. This means using renewable feedstock for plastic-based fibres and not using fossil-fuel-based fertilizers or pesticides in the farming of biologically-based input. A new textile economy further enables this shift to renewables as its very nature ensures that less energy and fewer resources are consumed.

4. A new textile economy regenerates natural systems and does not pollute the environment.

In a new textile economy where renewable resources are extracted from nature, this is carried out by regenerative and restorative methods that allow for the maintenance or improvement of soil quality and rebuild natural capital. In particular, this means using regenerative agriculture for biological-based input such as cotton, and sustainably managed forests and plantations for wood-based fibres. Substances of concern do not leak into the environment or risk the health of textile workers and clothing users. Plastic microfibres are not released into the environment and ocean. Other pollutants, such as greenhouse gases, are also designed out.

5. A new textile economy reflects the true cost (environmental and societal) of materials and production processes in the price of products.

In a new textile economy, the price of clothing reflects the full costs of its production, including environmental and societal externalities. Such costs are first analyzed and presented in company reporting, and ultimately reflected in product prices.

6. A new textile economy is distributive by design.

As part of promoting overall system health, a new textile economy presents new opportunities for distributed and inclusive growth. It creates a thriving ecosystem of enterprises from small to large, retaining and then circulating enough of the value created so that businesses and their employees can participate fully in the wider economy.

Words:

notwithstanding [ˌnɒtwɪθˈstændɪŋ]	prep. 尽管，虽然
linear [ˈlɪniə(r)]	adj. 线的；直线的；线状的
extract [ˈekstrækt]	v. 提取，提炼
landfill [ˈlændfɪl]	n. 废物填埋地
incinerate [ɪnˈsɪnəreɪt]	v. 焚毁，把……烧成灰烬
hazardous [ˈhæzədəs]	adj. 有危害的
microfibre [ˈmaɪkrəʊfaɪbə]	n. 微纤维
microbead [maɪkˈrəʊbiːd]	n. 微珠

inexorably [ɪnˈeksərəbli]	adv. 不可逆转地，不可阻挡地
catastrophic [ˌkætəˈstrɒfɪk]	adj. 灾难的；悲惨的；灾难性的，毁灭性的
disruption [dɪsˈrʌpʃn]	n. 扰乱，打乱，中断
restorative [rɪˈstɒrətɪv]	adj. 有助于复元的
ambition [æmˈbɪʃn]	n. 追求的目标；夙愿；抱负，雄心
inadvertently [ˌɪnədˈvɜːtntli]	adv. 无意地；不经意地
kick-start [ˈkɪk stɑːt]	vt. 促使……开始；使（项目）尽快启动
robust [rəʊˈbʌst]	adj. 强健的；强壮的；富有活力的
additive [ˈædətɪv]	n. （尤指食品的）添加剂，添加物
logistics [ləˈdʒɪstɪks]	n. 物流
converge [kənˈvɜːdʒ]	vt. 使汇聚 vi. 聚集；靠拢；收敛
functionality [ˌfʌŋkʃəˈnæləti]	n. 设计目的；设计功能，实用
sort [sɔːt]	vt. 整理；把……分类
feedstock [ˈfiːdstɒk]	n. （制造产品的）原料
offcut [ˈɒfkʌt]	n. （纸张、木材、布料等的）下脚料
externality [ˌekstɜːˈnæləti]	n. 外部效应

Critical reading and thinking

Task 1 Overview

There should be other ways to reduce textile pollution and boost green economy, work in pairs and find out more about those ways.

Task 2 Group discussion

Work in groups of 4-5 and have a discussion about the following questions.

1. What kind of pollution does textile industry cause to the environment?
2. Besides the way proposed in the text, what other ways can you suggest to reduce textile pollution?
3. What does green economy mean to the society and human beings?
4. Are there any more ways to achieve green economy?

Task 3 Language building-up

Translate the following terms from English into Chinese or vice versa.

take-make-dispose model	
regenerative agriculture	
industry-wide	
non-renewable resources	
化石燃料	
价值链	
逐步淘汰	
一次性的	

Task 4 Translation

Translate the following passage into Chinese.

A New Textile Economy Could Bring Substantial Benefits

In addition to offering benefits to business and the economy, a circular economy is beneficial to citizens and society, and it regenerates the environment. Research indicates that, given the global size and impact of the textiles sector, a new textile economy could play a major role in providing such benefits. A new textile economy would significantly lower the costs to businesses of using virgin materials. Decreased material use would also reduce businesses' exposure to volatile raw material prices and thereby increase their resilience. Realising these benefits for the textiles industry is dependent on radically increasing the amount of clothing that gets recycled by improving the current recycling system.

Additional profit opportunities for businesses through new services. Introducing new rental and subscription models allows businesses to build long-term customer relationships. These alternatives to the traditional sales model for clothes would allow fashion brands to create profits without having to increase throughput, and open up opportunities for innovators to trial new business models. In addition, value would be created through enhanced resale as well as by offering additional services before and during use, such as individualization, warranties, and maintenance.

Additional economic growth. A new textile economy means growing the most restorative and regenerative parts of the value chain, particularly those that make more productive use of material inputs (mainly through higher rates of clothing utilization and recycling of materials) and increase revenue from new circular activities. While some sectors (e. g. the production of virgin materials and certain clothing production activities) could expect reduced revenue, overall income would be expected to increase, which could boost economic growth.

Task 5 Research

Surf the Internet for more information about recyclable textile and green economy.
(You can try https://www.ftc.gov/tips-advice/business-center/guidance/, http://www.recycleaid.co.uk/cloth-and-fabric-recycling/.)

Task 6 Further reading

Global Textile Market to Register CAGR of 4.3% from 2020 to 2027

The global textile market size was valued at $961.5 billion in 2019 and is estimated to exhibit a CAGR of 4.3% from 2020 to 2027 owing to the increased demand for apparels, especially in developing countries such as China, India, Mexico, and Bangladesh, according to Grand View Research.

Furthermore, increasing disposable income and rapid urbanization has led to a rise in the number of supermarkets and retail stores, thereby driving the overall market growth.

Textile is a flexible material that is formed using numerous processes, including knitting, weaving, crocheting, or felting. These materials are extensively used to manufacture a wide range of finished goods, such as upholstery, kitchen, transportation, bedding, construction, medical, protective equipment, apparel, handbags, and clothing accessories.

The U.S. is expected to be the largest market for textiles in the North American region. Textile companies in the region focus on restructuring their businesses, developing effective work processes, and investing in niche products. Natural fibres are anticipated to be the largest product segment in the region on account of the rising demand from the fashion and apparel industry.

There has been increasing awareness about personal protective equipment (PPE) in the manufacturing industry on account of stringent regulations for worker safety. This is expected to lead to an increased demand for engineered fibre products such as nylon. Furthermore, technological innovations in terms of the development of new upholstery products derived from coated fabrics and spider silk are expected to open new industry avenues over the forecast period.

The rising popularity of sophisticated gadgets with varied technologically-advanced functions such as sensing and reacting to surroundings is anticipated to drive the demand for smart textiles over the forecast period. Smart textiles are used by consumers as a clothing entity as well as by military professionals for protection and safety purposes.

Waterproof breathable textiles (WBT) are durable, lightweight, and deliver high strength. As a result, they are extensively used in numerous outdoor sports products including tents, backpacks, footwear, and other fabric-based gears or garments. Increasing demand for lightweight and multifunctional fabric in the sportswear sector is likely to boost the demand for waterproof breathable textiles.

Based on the raw materials, the market can be segmented into cotton, wool, silk and others. In terms of volume, cotton is anticipated to be the largest raw material segment, accounting for a market of 39.5% in 2019. The growth of the segment is attributed to the properties of cotton including high absorbency and strength, and color retention. China, India, and the U. S. are the major cotton producers in the world.

Chemicals play an important role in the textile industry. Acetic acid, oxalic acid, sulfuric acid, and soda ash are some of the basic chemicals during textile manufacturing. Chemicals are used as finishing and dyeing agents in order to improve the appearance of textiles. The rising importance of the physical appearance of textiles is expected to drive the segment growth over the forecast period.

Wool-based textile accounted for 12.6% of the market in terms of volume in 2019. Wool fibre, which is also known as keratin, is composed of hydrogen, nitrogen, carbon, and sulfur. Wool is mainly used in the manufacturing of winter wear. It is recyclable and renewable and also has qualities of strength and moisture management, thereby driving the product demand.

Silk is increasingly adopted in the textile industry owing to properties such as high strength, elasticity, and absorbency. In terms of revenue, this segment is projected to register the highest CAGR of 4.7% from 2020 to 2027.

Based on product, the textile market has been segmented into natural fibres, polyesters, nylon, and others. The natural fibres segment dominated the market in 2019 owing to the growing demand from the apparel and fashion industry. Natural fibres include cotton, linen, flax, silk, hemp, and wool. They are considered to be biodegradable and renewable and are hence eco-friendly.

Polyester fibre is characterized by high strength, quick-drying, and high chemical and wrinkle-resistant properties. These advantages have made polyester one of the most widely used products in the textile industry. Polyester finds application in a variety of end-use products such as carpets, curtains, nets, and ropes. Growing demand for household products and technical textiles is projected to drive the polyester segment over the forecast period.

Nylon is widely used in the synthetic textile category. It exhibits properties such as high resilience, elasticity, luster, and low moisture absorbency, and hence finds application in several industries including apparel and home furnishing. Since nylon is a highly versatile material, it is widely used in the textile industry.

Others include polyethylene (PE), polypropylene (PP), aramid, and polyamide. Polypropylene is a synthetic fibre that has high strength and water-resistant properties. Similarly, polyamide is known for its high strength and durability. Due to these factors, these fibres are widely used in the textile industry.

On the basis of application, the market has been segmented into household, technical, fashion and clothing, and others. Fashion and clothing are expected to be the fastest-growing application segment over the forecast period owing to the rise in consumer spending on apparel and clothing, coupled with the highly developed fashion industry in the European and North American regions.

Fashion and clothing is the largest consumer of textiles. In terms of volume, the fashion sector held a considerable share of over 70.0% of the total market in 2019. Apparel, ties and clothing accessories, as well as handbags are the key areas that significantly consume textiles. Consumer requirements for crease-free fabrics and high quality dyed and printed fabrics are expected to drive the demand for textiles over the forecast period.

Textiles are utilized in different areas of household applications, some of which include bedding, kitchen products, upholstery, and towels. Natural fibres such as cotton and linen as well as synthetic fibres including polyester and acrylic polyamides are primarily used for the manufacturing of household textiles. Rising consumer preference for light absorbing and textured fabrics in home decor is anticipated to drive the segment growth over the forecast period.

Technical textiles are used to meet the high-performance end-user requirements from various industries including construction, transportation, medical, and protective equipment. Nylon, polypropylene, polyester, and acrylic fibres are some of the widely used raw materials used for manufacturing technical textiles.

Asia Pacific is the largest regional market and is anticipated to register a substantial CAGR of 5.6% in terms of value over the forecast period. This is attributed to the rapidly increasing demand for apparels, particularly through e-commerce portals. Moreover, manufacturers prefer setting up manufacturing units in countries such as China, India, Bangladesh, and Pakistan owing to high cotton production and low labor costs.

Europe was the second-largest market in terms of value as well as volume in 2019. Strict regulations imposed by the European Union (EU) on product quality and distribution of fabrics are anticipated to stabilize the growth rate in the region over the forecast period. Demand for silk fabrics for household and fashion applications is considered to provide substantial scope for market growth in Europe.

Easy availability of raw materials such as cotton and polyester, coupled with the presence of textile manufacturing units, has made Central and South America one of the largest producers of textiles. Countries such as Venezuela, Nicaragua, and Brazil are expected to drive the regional market for textiles over the forecast period on account of increasing construction spending, free trade agreements, and abundant availability of raw materials.

Chapter 6　Technical Textiles and Smart Fabrics

Section 1　General Introduction to Technical Textiles

Lead-in

Warm-up questions:

1. What roles do you think textiles can play in production besides those in our daily life?
2. What is the function of mechanical lungs for patients with breathing problems?

Lexical chunk bank	
技术纺织品	technical textiles
医用纺织品	medical textiles
防护装备	protective equipment
服装纺织品	apparel textiles
防弹背心	bulletproof vest
滤布	filter clothing
防晒材料	sunscreen materials
防护材料	proofing materials
纺织建筑学	textile architecture
透水织物	permeable fabrics
聚合物植入物	polymeric implants
隔离服	isolation gowns
外科手术服	surgical clothing

Section 1 General Introduction to Technical Textiles

压缩服	compression garment
工业纺织品	industrial textiles
针状毡	needle felts
阻隔材料	barrier materials
包装织物	packaging textiles
汽车内饰	interior of cars
运动服装	sports clothing

1. Introduction to technical textiles

A **technical textile** is a textile product manufactured for non-**aesthetic** purposes, whose function is the primary criterion (标准). Technical textiles mainly involve several applications such as **medical textiles**, **protective equipment** (e. g. heat and **radiation** protection for fire fighter clothing, molten metal protection for welders, stab protection and **bulletproof vests**, and spacesuits) and so forth.

The sector is large, growing, and supports a vast **array** of other industries. The global growth rate of technical textiles is about 4% per year, which is greater than that of home and **apparel textiles**, which are growing at a rate of 1% per year. Currently, technical textile materials are most widely used in filter clothing, furniture, hygiene medicals and construction material.

2. Classification

Technical textiles can be divided into many categories, depending on their end use. Technical textiles **specify** 12 application areas: Agrotex, Buildtex, Clothtex, Geotex, Hometex, Indutex, Medtex, Mobiltex, Oekotex (Ecotex), Packtex, Protex and Sportex.

Agrotech (**Agro-textiles**)

Agro-textiles are applied in the agrotech sector aimed at crop protection and crop development and reducing the risks of farming practices. Primarily agro-textiles offer **weather resistance** and the resistance to microorganisms and protection from unwanted elements and external factors. Agro-textiles help to improve the overall conditions with which crop can develop and be protected. There are various textile products, fabric forms, fibres and techniques used in agro-textiles. The examples include shade nets(遮荫网), **thermal insulation** and **sunscreen materials**, windshield, antibird nets, which provide minimal shading and proper temperature as well as air circulation for protecting plants from direct sunlight and birds. Agrotextiles involve **hail** protection nets, and crop covers, etc. Agro-textiles are useful in horticulture (园艺学), aquaculture (水产养殖), **landscape**

gardening and forestry also. More examples of use and application cover **livestock** protection, **suppressing** weed and insect control, etc.

Buildtech (Construction textiles)

Construction textiles are used in construction of concrete reinforcement, **facade foundation** systems, interior construction, insulations, **proofing materials**, air conditioning, noise prevention, **visual** protection, protection against the sun and building safety.

An interesting and aesthetic appealing application is the use of textile membranes(膜) for roof construction. This area is also referred to as **textile architecture**. PVC coated high tenacity PES(高韧性聚醚砜), teflon coated glass(聚四氟乙烯涂层玻璃) fibre fabrics or silicone coated PES(有机硅涂层聚醚砜) are used for their low creep properties(蠕变性能). Splendid examples of such construction are found in football stadiums, airports and hotels.

Geotech (Geotextiles)

They are used in reinforcement of **embankments** or in construction work. The fabrics in geo textiles are **permeable fabrics** and are used with soils having the ability to separate, filter, protect or drain. The application areas include **civil engineering**, earth and road construction, dam engineering, soil sealing and in **drainage systems**. The fabrics used in this area must have good strength, **durability**, low **moisture absorption** and thickness. Mostly non-woven and woven fabrics are used in it. Synthetic fibres like glass, polypropylene (聚丙烯) and acrylic fibres (丙烯酸纤维) are used to prevent **cracking** of the concrete, plastic and other building materials. Polypropylene and polyester are used in geotextiles and dry/liquid filtration due to their **compatibility**.

Indutech (Industrial textiles)

They are textiles used for chemical and electrical applications and textiles related to **mechanical engineering**, silk-screen printing, filtration, plasma screens(等离子显示屏), propulsion (推进力) technology, lifting/conveying equipment, **sound-proofing elements**, melting processes, roller covers, grinding technology, insulation, seals and fuel cell.

Textiles for transports

The idea of using textiles to transport humans or goods is not recent. Textile ropes and sails have been **instrumental** in powering ships since the early days of human civilization and the birth of human **aviation** in the form of balloons and early airplane **prototype**s is equally textile linked.

The modern concept of mobility enabling textiles comes in the forms of:

➢ performance fibre-based textiles used in balloons, parachutes(降落伞), sails, nets and ropes;

➢ aircraft wing and body structures or boat rumps (船舷坡道) made of fibre and textile-

based composites;
- inflatable (可膨胀的) components of satellites or other spacecraft;
- flexible **reservoirs**, containers or bags used for transportation of gases, liquids and bulk goods by road, rail, water or air.

Medtech (Medical textiles)

Medical textiles are the textile materials such as fibres, yarns and fabrics that support medtech (area of application) with healthcare, **hygiene**, infection control, **barrier materials**, **polymeric implants**, medical devices and smart technologies. The medical textiles help with variety of products in handling medical practices and procedures such as treating an injury and dealing with a medical environment or situation. Medical textiles also include fibres for growing human organic tissues.

Medical textiles offer laminated and coated material (层压和涂层材料) for various gowns for better protection from infections or fluids, such as PPE gowns for doctors, nurses, hospital staff and gowns worn by healthcare personnel as personal protective equipment, patient gowns and surgical and **isolation gowns**. The technical textiles for medical use also help in providing facemasks, **surgical clothing** and drapes, disposable nitrile gloves(一次性腈手套) (**goggles**, head caps, long shoe covers hoods, various wound care assistance such as bandages and dressings, and **compression garments**, etc. Medical textiles integrated with smart technologies provide remote contact between doctor and a user.

In the United States, medical gowns are medical devices regulated by the FDA. Surgical isolation gowns are regulated by the FDA as a Class II medical device. Surgical gowns only require protection of the front of the body due to the controlled nature of surgical procedures, while surgical isolation gowns and non-surgical gowns require protection over nearly the entire gown.

Mobiltech (Textiles used in transport; automotive and aerospace)

These textiles are used in the construction of automobiles, railways, ships, aircraft and spacecraft. Examples are truck covers (PVC coated PES fabrics), car trunk coverings (often needle felts), lashing belts for cargo tie downs (货物系带), seat covers (knitted materials), seat belts, nonwovens for cabin air filtration (also covered in indutech), airbags, parachutes, boats (inflatable), air balloons.

Many coated and reinforced textiles are used in materials for engines such as air ducts (导管), timing belts, air filters, non-wovens for **engine sound isolation**. A number of materials are also used in the **interior of cars**. The most obvious are seat covers, safety belts and airbags but one can find textiles also for the sealing. Nylon gives strength and its bursting strength is high, so it is used as air bags in cars.

Carbon composites (合成物) are mostly used in the manufacture of aeroplane parts while carbon fibre is used for making higher end tires.

Oekotech or Ecotech (Ecological Protection textile)

Ecotech textiles are those for environmental protection-floor sealing, **erosion** protection, air cleaning, prevention of water pollution, water cleaning, waste treatment/recycling, depositing area construction, product extraction and **domestic water sewerage plants**.

Packtech (**Packaging textiles**)

Packaging textiles are the materials such as fibres, yarns, fabrics, and polymer(聚合物) which contribute to manufacture various packaging, containers, bags, **lashing straps**, canvas covers, marquee tents(大帐篷) and so forth.

Protech (**Protective textiles**)

The main target of the technical protective textiles is to improve people's safety in their workplaces. A technical protective fabric can save a worker's life. That's why most of them are mainly used to manufacture PPE (personal protective equipment). The demand of these fabrics is growing around the world as a result of people's requirement for more safety at work. The aim of a technical protective fabric isn't fashion. They are designed to have extra values in protection against some hazards.

Nowadays protective textiles can be used to protect people from the following:
- high temperatures (insulating, firefighters)
- burns (flame, convective (对流的) and radiant(辐射的) heat, firefighters, ATEX area)
- electric arc flash discharge(电弧闪光放电) (plasma explosion(等离子体爆炸), electric companies)
- molten metal impacts (foundries)
- metal sparks (welding)
- **acid** environment (petrochemical(石油化学制品), gas, refineries, chemical)
- bullet impact (military, security)
- cut resistant (gloves, glass industry)
- astronaut's suits
- **leftover** food packets

Sportech (**Sports textiles**)

Sports and fitness are other popular areas where technical textiles are promising. At present, **user comfort and safety** are the primary goals. Garments made with smart textiles can **incorporate** several functions, including anti-chafing (防皮肤发炎) and heating elements. We can also find sports clothing that include vital **statistic monitoring**, including heart rate, breathing, body temperature, steps taken, muscles stimulated and more. This record of performance can help you track achievements and protect against injury.

Technical textiles can also be used to detect head injury in **real-time**. Some protective equipment, such as mouth guards and helmets, are equipped with sensors to identify **bruising** and **anomalies** in blood flow. When an athlete is injured, paramedics(护理员) will already have the data and can determine the best possible treatment right away.

Section 1　General Introduction to Technical Textiles

Words:

aesthetic [iːsˈθetɪk]	adj. 美学的；审美的；有美感的
specify [ˈspesɪfaɪ]	vt. 详细说明；指定；阐述
radiation [ˌreɪdiˈeɪʃn]	n. 辐射；放射线
array [əˈreɪ]	n. 一系列；大批；数组；陈列
hail [heɪl]	n. 冰雹
livestock [ˈlaɪvstɒk]	n. 家畜；牲畜
suppress [səˈpres]	vt. 镇压；隐瞒；压制；止住；禁止
visual [ˈvɪʒuəl]	adj. 视觉的；视力的；看得见的；形象的
embankment [imˈbæŋkmənt]	n. 堤防；路基；堤岸；(铁路的)路堤
drainage [ˈdreɪnɪdʒ]	n. 排水；排水系统；污水
durability [ˌdjʊərəˈbɪləti]	n. [U] 持久性；耐久性
crack [kræk]	vi. 破裂；砸
compatibility [kəmˌpætəˈbɪləti]	n. 和谐共处；协调；兼容
instrumental [ˌɪnstrəˈmentl]	adj. 乐器的；有帮助的
aviation [ˌeɪviˈeɪʃn]	n. 航空；飞机制造业
prototype [ˈprəʊtətaɪp]	n. 原型；范例；雏形
reservoir [ˈrezəvwɑː(r)]	n. 水库；储藏；蓄水池；积蓄
hygiene [ˈhaɪdʒiːn]	n. 卫生；卫生学
goggle [ˈgɒgl]	n. 眼睛睁视；护目镜
bandage [ˈbændɪdʒ]	n. 绷带
composite [ˈkɒmpəzɪt]	n. 合成物；复合材料
erosion [ɪˈrəʊʒn]	n. 腐蚀；侵蚀；流失
sewerage [ˈsuːərɪdʒ]	n. 排水设备；污水
acid [ˈæsɪd]	n. 酸；酸性物质
leftover [ˈleftəʊvə(r)]	n. 剩饭；残留物
proof [pruːf]	vt. 检验；给……提供防护措施

Chapter 6 Technical Textiles and Smart Fabrics

bruising ['bruːzɪŋ]	n. 挫伤
incorporate [ɪnˈkɔːpəreɪt]	vt. 合并
anomaly [əˈnɒməli]	n. 反常之事物；不规则

Phrases:

non-aesthetic purposes	非审美目的
hygiene medicals	保健药品
weather resistance	耐候性
ecological protection	生态保护
thermal insulation	隔热
landscape gardening	园林绿化
facade foundation	立面基础
civil engineering	土木工程
drainage system	排水系统
moisture absorption	吸湿
mechanical engineering	机械工程
sound-proofing element	隔音元素
bursting strength	爆裂强度
grinding technology	研磨技术
human organic tissue	人体器官组织
engine sound isolation	发动机隔音
domestic water sewerage plants	生活污水处理厂
lashing straps	系带
high tenacity	高韧性
user comfort and safety	用户舒适度和安全度
statistic monitoring	数据监控
real-time	实时的

Section 1 General Introduction to Technical Textiles

Critical reading and thinking

Task 1 Overview

Complete the following notes about the text.

Application types	Definition	Application fields	Examples
Agrotex			
Buildtex			
Clothtex			
Geotex			
Hometex			
Indutex			
Medtex			
Mobiltex			
Oekotex			
Packtex			
Protex			
Sportex			

Task 2 Pair work

Work in pairs and retell the text with the information in Task 1. Use as many lexical chunks as possible.

Task 3 Group discussion

Work in groups of 4-5 and have a discussion about the following questions.

1. Which areas of the 12 application areas of technical textiles do you think occupy the largest market shares of textile industry? List the top 3 areas and state the reasons.
2. Which area of technical textiles or combination of different areas can be applied in the fight against COVID-19? How?

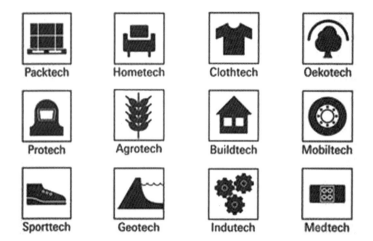

Task 4 Language building-up

I. Translate the following terms from English into Chinese or vice versa.

apparel textiles	
geo textiles	
industrial textiles	
synthetic fibres	
Agrotech (Agro-textiles)	
防护纺织品	
技术纺织品	
隔离服	
外科手术服	
用户舒适度和安全度	

II. Complete the following sentences with the words or phrases given in the previous exercise.

1. _____ is a textile product manufactured for non-aesthetic purposes, whose function is the primary criterion.

2. _____ are textiles used for chemical and electrical applications and textiles related to

mechanical engineering.

3. The fabric used in _____ must have good strength, durability, low moisture absorption and thickness.

4. At present, _____ are the primary goals of sports textiles.

5. _____ are applied in the agrotech sector aimed at crop protection and crop development and reducing the risks of farming practices.

Task 5 Guessing game

Guess the missing lexical chunks from different contexts with the ones in the box.

A. knitted material	B. surgical gown	C. medical textiles

Group 1

1. I recall the greenness of the _____ and the thick fingers of the resident as she inserted a needle into the top of my hand.

2. Now, swaddled in blankets in the cold operating room and wide awake, Joan looked up at half a dozen physicians in _____, all of whom seemed to be shouting orders at her simultaneously.

3. Three years ago, he opened a London restaurant called Pharmacy, decorating it with pill-shaped bar stools, medicine cabinets and, in the men's room, a collage of hospital detritus piled behind a glass wall. Waiters serve drinks in _____.

Group 2

1. In the field of textile hygiene, we can check the barrier effect of _____ and personal protection equipment, the microbiological bacterial contamination of feather/down and washing products and the effectiveness of disinfectant washing processes.

2. These spun fibres promise to address the manufacturing challenges in fields as diverse as artificial organ creation, drug delivery systems, advanced filtration media, and _____.

Group 3

1. The very latest addition is evening dresses in the finest _____.

2. Processing of very fine yarns, in machines with gauges of E40 and finer, enables the creation of very light _____.

3. The medium ball point FG is a versatile point style, yet is especially suited to _____.

Task 6 Translation

Translate the following paragraph into Chinese.

During the COVID-19 pandemic, health authorities in some counties have recommended that citizens wear masks in public under certain circumstances. In this context, a number of grassroots initiatives have emerged to help people sew cloth masks at home for their personal use and in some communities to supply nearby hospitals. These improvised masks typically overlook some of the design elements that were crucial for the efficiency of earlier cotton masks. Yet the public response has been enthusiastic in some places, at least as measured by the number of people viewing instructional videos. The home production of reusable masks for use in the community offers last resort solutions to some and comfort to many, but is unlikely to contribute more than marginally to solving the shortage of personal protective equipment globally. As for health-care workers and hospitals, in some settings, they are experimenting with methods to sanitize disposable masks, even though they were not designed to be reused. Such an approach is a far cry from the carefully designed, manufactured, and tested reusable masks in use in the 1970s.

Task 7 Research

Surf the Internet for more information about technical textiles and their segments.

(You can try http://www.ittaindia.org/?q=abouttechnicaltextile, http://www.globaltextiles.com and http://www.innovationintextiles.com.)

Step 1 Look for lexical chunks you think worth learning and share the new findings to the class.

Step 2 Work in group of 4-5. Each group will be responsible for a presentation of a technical textile segment to the class.

TECHNICAL TEXTILE SEGMENTS

- AGROTECH (AGRO TEXTILES)
- BUILDTECH (CONSTRUCTION TEXTILES)
- CLOTHTECH (CLOTHING TEXTILES)
- GEOTECH (GEO TEXTILES)
- HOMETECH (HOME TEXTILES)
- INDUTECH (INDUSTRIAL TEXTILES)
- MEDITECH (MEDICAL TEXTILES)
- MOBILTECH (AUTOMOTIVE TEXTILES)
- PACKTECH (PACKAGING TEXTILES)
- PROTECH (PROTECTIVE TEXTILES)
- SPORTECH (SPORTS TEXTILES)

Section 2 Detailed Introduction to Some Technical Textiles

Lead-in

Warm-up questions:

1. What roles can medical textiles play in an operation room?
2. What dangers do those firefighters face in action? And what protective functions should their fighting suit have?

Lexical chunk bank	
生物危害防护服	biohazard suit
高性能纺织材料	high performance textile materials
防渗材料	impermeable materials
耐久弹道织物	durable ballistic fabric
防冲击织物	impact protection fabrics
防紫外线服装	ultraviolet-resistant clothing
绝缘材料	insulation materials
织物专制力	fabric imperious
耐温服装	temperature-resistant clothing
高级爆炸物处理炸弹套装	advanced EOD bomb suit
低温(织物)	microtherm
合成材料	synthetic materials
非植入性材料	non-implantable materials
中空粘胶	hollow viscose
植入性医用纺织品	implantable medical textiles
石膏布	plaster fabrics

纤维复合材料	fibre composites
车顶衬里地毯	headliner carpets

1. Protective Textiles

One of the most **prominent** applications of technical textiles is to prevent bodily injury and harm. In general, this type of clothing focuses on four areas—**contamination**, impact, temperature and ultraviolet.

In industrial and disaster areas, workers often operate in **toxic** environments. A **biohazard suit** made from **impermeable materials** can prevent contamination. Certain **impact protection fabrics**, typically used by police and armed forces, can counter the risk of an explosion, shooting, stabbing or **blunt impact.** The once-popular material, Kevlar（芳纶）, has now been replaced by a **durable ballistic fabric** made of HPPE（high-performance polyethylene, 高性能聚乙烯）. The field of **temperature-resistant clothing** has recently seen rapid innovation. Nomex, made by Du Pont, is a material that can withstand temperatures of up to 400 degrees Celsius without charring. Similar advancements have been made in **ultraviolet-resistant clothing**, mainly swimwear.

Protective clothing types

Protective clothing can be divided into the following types:

— Clothing against heat and flame
— Clothing against mechanical impacts
— Firemen's protective clothing
— Clothing against cold
— Clothing against **foul** weather（moisture wind）
— Clothing against chemical substances（gas liquids, **particles**）
— Clothing against **radioactive contamination**

Biological protection

Biological protection comes in two categories. The first is the protection of humans and earth from harm. It usually involves stopping disease（or other kind of damages）from either natural sources or human caused accidents. The second is protection of potential new alien life forms from harm. It involves stopping humans from damaging alien life in space.

Chemical protection

Protective textiles used for chemical production are categorized from high to low. They contain electrometric barriers（电位势垒）such as butyl rubber（丁基橡胶）and provide excellent protection from **chemical warfare agents**, but wearers can use the clothing only intermittently（断断续续地）due to heat stress, motion restriction and weight. These fabric systems are heavy and subject the soldiers to heat stress under high workload and battlefield conditions.

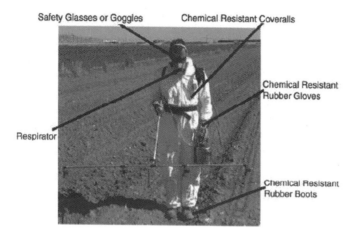

Advanced EOD bomb suit(高级爆炸物处理炸弹套装) and helmet

An advanced design bomb suit and helmet that offers highest ballistic protection in the world is shown in the picture below. The suit is constructed from Kevlar with an outer antistatic cover of Nomex/Kevlar and comprises of a jacket, crotchless trousers(无叉裤), groin cup and **rigid ballistic panels**. The suit itself is light in comparison with other suits, with **front protection plates** and that reduces operator's fatigue.

Heat protective

Microtherm has been one of the leading high temperature thermal insulation materials. It is a better **insulation material** even than still air and at high temperatures it has a **thermal conductivity** as much as four times lower than most conventional insulation ones such as fibre. **Microtherm** products cover a comprehensive range in both rigid and flexible forms, facilitating neat and effective design solutions to the many diverse thermal problems in military equipment.

Applications of smart textiles for protective purposes

There are various types of application of smart textiles for protective purposes. This kind of textiles is able to detect conditions that signal increased danger and prevent accidents by sending out a warning when hazardous conditions have been detected. In the case of serious threats, they can react by providing **instantaneous** protection. Apart from obvious threats like heat, chemicals, gases, etc., danger can also be caused by people themselves. Individuals can be threatened by an **acute** disease such as a heart attack, a stroke, or other physical conditions and may be unable to perform their tasks safely. For instance, people driving a machine or a vehicle must maintain a high and continuous level of concentration and awareness. Fatigue, consumption of alcohol or medication can negatively affect such parametres, leading to an increased risk of accidents. Ultimately the suit may communicate to the machine that the driver is no longer able to continue his operations safely and the machine may be stopped. One of the simplest reactions is color change, providing straightforward visual signals to both the wearer and the environment, so that adequate measures can be taken in times.

2. Medical Textiles

Medical textile is an emerging area with numerous uses. Medical textile products are produced from **high performance textile material**s for their functional and performance properties. Studies have carried out to impart the properties of wound care and **antibacterial finishes** in medical textiles.

Implantable medical textiles

The materials are used in effecting repair to the body whether it is wound closure(闭合) (sutures) or replacements surgery like vascular grafts(血管移植物), artificial ligaments(韧带), etc. Filament and texture yarns are used and coated to prevent **leakage** of blood while tissue is forming on the inner walls. Carbon fibre is a popular material for tissue repair. Suspensors and reinforcing surgical meshes are used in **plastic surgery** for repairing defects of the abdominal(腹部的) wall or other surgical treatments.

Non-implantable materials

These materials are used for external applications on the body and may or may not contact with skin. They are employed as covering, **absorbent**, protective and supports for injured or diseased part. There are different types of non-implantable materials.

The first one is primary wound dressings, which are placed next to the wound surface. The next one is absorbent, which is similar to wound pads used in surgery, manufactured from well-bleached, carded and cleaned cotton fabrics.

Extra corporeal devices

Extra corporal devices are **mechanical organs** that are used for blood **purification** and include the artificial kidney, the artificial liver, the artificial heart and the mechanical lung.

Artificial kidney

It is a tiny instrument, about the size of a two-cell flashlight, made with hollow hair-sized cellulose fibres (纤维素纤维) or hollow polyester fibre slightly latest than capillary vessels (毛细血管). Artificial kidney is used to remove waste products from patient's blood.

Artificial liver is made with **hollow viscose** (粘胶) to separate and dispose patient's plasma (血浆) and supply fresh plasma.

Artificial heart

The heart-lung machine is an 8-ounce plastic pump lined with dacom velourdacom velour (科姆丝绒) that maintains a patient's blood circulation and oxygenation during heart surgery by diverting blood from the venous (静脉的) system, directing it through tubing into an artificial lung (oxygenator), and returning it to the body. The oxygenator removes carbon dioxide and adds oxygen to the blood that is pumped into the arterial (动脉的) system. The blood pumped back into the patient's arteries is sufficient to maintain life at even the most distant parts of the body as well as in those organs with the greatest requirements (e.g., brain, kidneys, and liver). While the heart is relieved of its pumping duties, it can be stopped, and the surgeon can perform open-heart surgery that may include **valve repair** or replacement or repair of **defects** inside the heart.

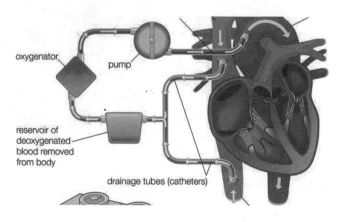

Mechanical lung

It is made with hollow polypropylene fibre (聚丙烯纤维) or a hollow silicone membrane(中空硅树脂膜) and is used to remove carbon dioxide from patient's blood and supply fresh oxygen.

Products used for medical surgical dressings

The modern wound dressing is usually made of three layers:

a. Wound contact layer: It should not stick to the wound or cause maceration (浸润) of the skin if the dressing is not changed. It can be woven, knitted or non-woven made from silk, viscose or polyethylene.

b. Middle **absorbing layer**: It is used to absorb blood or liquids while providing a cushioning effect to protect the wound. It is generally a non-woven textile composed of cotton or viscose.

c. Base material: It provides a means by which the dressing is applied to the wound. The material is coated with acrylic adhesive (丙烯酸类胶粘剂) to hold the dressing in place, eliminating the need for bandage.

Bandages

Bandages are designed to perform a whole variety of specific functions depending upon the final medical requirement. They can be woven, knitted, non-woven or composite in structure. They can be classified into the following several groups depending on their functions, simple bandages, **light support bandages**, **compression bandages** and so on.

Orthopedic bandages

These bandages are used under plaster casts and **compression bandages** to provide padding and prevent discomfort. Non-woven orthopedic cushion bandages (无纺骨科垫子绷带) are made from polyester(聚酯) or polypropylene and blends of natural and synthetic fibres. Light needle punching can **guarantee** greater cushioning effect.

Plaster

Plasters are made up of three layers: plaster fabrics, adhesive and wound pad. The modern plaster fabric is made from spun bonded nonwovens of cotton (纺粘非织造布), viscose, polyester or glass fibre. The adhesive used for **plaster fabric** is acrylic that doesn't stick to the skin.

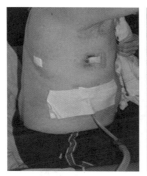 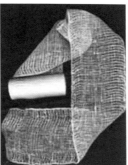

3. Automotive textiles

The growth of automotive textiles is rapid during the past decades. There are a few driving forces behind the growth of automotive textiles. These forces could be improvement in the standard of living of people resulting in the greater demand for personal vehicles. A car interior has become more and more important as people are spending more time in cars. For better fuel economy, the trend is towards light weight vehicles with **fibre composites** in most of the applications.

Ecological reforms for recycling of used cars have increased the amount of textiles in an automobile. Apart from interiors and safety, textiles have also come up with the solution to engineering problems such as tire reinforcement, acoustics (声音) protection as well as gas and air **filtration**.

Car interior

People are spending more time in car due to increased traffic density, greater mobility and long-distance traveling. Car interiors are becoming increasingly important with higher consumer expectations. Car interiors comfort is an important priority reflected in their costs of car interiors.

Today, over 90% car seat covers are made of polyester filament yarns. Apart from good abrasion resistance(耐磨性), polyester also offers good tearing strength, ease of cleaning and mildew resistance (防霉性). Wool is sometimes used due to its flame resistant characteristic but it is only limited to upper end cars.

Seat belts

Seat belt is an energy absorbing device, which controls the forward movement of wearer in an event of sudden deceleration of vehicle. Seat belt is designed to keep the load **imposed** on victim's body during crash down to survivable limits and deliver non-recoverable extension in an event of crash. Wearing of seat belt can reduce fatal and serious injuries by 50%. Seat belts are designed to hold the occupants in correct position to strike air bag when it is inflated. Thus in modern cars, seat belts and the air bags are not substituted to each other but **complementary**.

Polyester is the most preferred fibre for seat belts as it satisfies the requirements of **maximum** extension up to 24% to 30% and because of its good abrasion resistance, heat and light resistance and light weight. About 90% seat belts are made from polyester only.

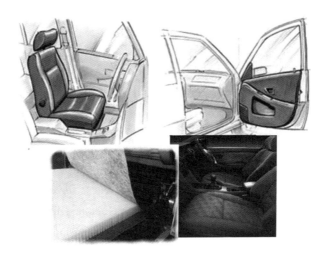

Air bags

In most countries air bags are **mandatory** for all passenger cars due to stringent(严厉的) legislation. According to the report, air bags system has contributed up to 20% reduction in casualties resulting from front **collision**. Air bags cushion an occupant in an event of crash, which helps avoid the heat or collision.

The working of air bag is precision application. In just 0.03 seconds air bags should begin and by 0.06 seconds after crash, the bag should be fully inflated. Air bags may be built into **steering** or in some other strategic location. Fabrics used for air bags must be able to withstand force of hot gases and they must not **penetrate** through fabric.

Carpets

There is about 3.5 to 4.5 square meters carpet in each car. Apart from ethical and **sensual** comfort, carpets also play significant roles in acoustic and **vibration** control. Increasing popularity of multi-purpose vehicles and **headliner carpets** had also increased the demand of the same.

Road noise is considered as an environmental pollution in few countries. There are pressures on automobiles to reduce external noise about 50%. Carpets are contributing to solving that problem. Carpets, by providing thermal and acoustic protection, directly contribute to safety.

Tire cord fabrics(帘子布)

Tire cord fabric is skeleton structure, which holds the uniform rubber **mass** of tire. It gives **dimensional** stability to tire. Tire is a pressure vessel and cord fabric keeps it dimensionally stable. Cord fabric also gives load carrying capacity to tires.

Here is a conclusion of the fibres used in automobiles:

Application	Fibres used
Seat covers	Nylon, polyester, pp, wool
Seat belt	Polyester
Carpet	Nylon, PET, PP
Air bags	Nylon 66, Nylon 46
Tyre cords	Viscose rayon, nylon, Kevlar
Composites	Carbo, glass, armid

Words:

prominent [ˈprɔminənt]	adj. 明显的；著名的；杰出的；广为人知的
contamination [kənˌtæmiˈneiʃn]	n. 污染；污秽；污物；混淆不清
toxic [ˈtɔksik]	adj. 有毒的；中毒的；有害的；致命的；恶毒的
foul [faul]	adj. 污浊的；恶臭的；险恶的；(言辞)粗俗的
particle [ˈpɑːtikl]	n. 微粒；极少量；质点，粒子
ballistic [bəˈlistik]	adj. 弹道(学)的；飞行物体的
facilitate [fəˈsiliteit]	vt. 使容易；使便利；促进；帮助
instantaneous [ˌinstənˈteinjəs]	adj. 立刻的，瞬间的；即时的
acute [əˈkjuːt]	adj. 敏感的；敏锐的；剧烈的
medication [ˌmediˈkeiʃn]	n. 药疗法；加入药物；药物处理；医药，药物
parametre [pəˈræmitə]	n. 通径；参数；变数；(结晶)半晶轴
bleach [bliːtʃ]	n. 漂白剂；漂白 v. 变白；漂白
purification [ˌpjuərifiˈkeiʃn]	n. 净化；提纯
leakage [ˈliːkidʒ]	n. 漏；泄漏

Chapter 6 Technical Textiles and Smart Fabrics

defect [diˈfekt]	n. 缺陷；过失；缺点；弱点，瑕疵
absorbent [əbˈsɔːbənt]	adj. 能吸收的；有吸收力的 n. 吸收剂；中和剂
compress [kəmˈpres]	vt. 压缩；压榨；镇压
guarantee [ˌɡærənˈtiː]	n. 保证(人)；保证书 vt. 保证；承诺；确定；保护
filtration [filˈtreɪʃn]	n. 过滤；滤清
impose [imˈpəuz]	vt. 加(负担、惩罚等)于；课(税)；处(罚)
complementary [ˌkɔmpliˈmentəri]	adj. 补足(充)的
maximum [ˈmæksiməm]	n. 最大量；最大值 adj. 最高(大)的；最大值的
mandatory [ˈmændətəri]	adj. 法定的；义务的；强制性的 n. 受托管理者
collision [kəˈliʒn]	n. 猛烈相撞；抵触；碰撞；互撞
steer [stiə]	vi. 驾驶汽车；操舵
penetrate [ˈpenitreit]	vt. 穿过；透入；浸透；了解；进入
sensual [ˈsensjuəl]	adj. 感官的；肉体的
vibration [vaiˈbreiʃn]	n. 振动；动摇；不安
mass [mæs]	n. 团，质量，大量
dimensional [dɪˈmenʃənəl]	adj. 维度的；次元的；空间的；尺寸的

Phrases：

radioactive contamination	放射性污染
blunt impact	钝性冲击
chemical warfare agents	化学战剂
rigid ballistic panels	刚性弹道板
front protection plate	前护板
thermal conductivity	导热系数
reinforcing surgical meshes	加固手术网
artificial(kidney, liver, heart)	人工(肾脏、肝脏、心脏)
replacements surgery	置换手术

antibacterial finishes	抗菌整理
absorbing layer	吸收层
plastic surgery	整容手术
extra corporeal device	体外装置
mechanical organ	机械器官
valve repair	（心脏）瓣膜修复
light support bandage	轻型支撑绷带
compression bandage	压缩绷带
plaster cast	石膏铸件
plaster fabrics	石膏织物
light needle punching	轻型针刺
fuel economy	燃油经济性
ecological reform	生态改革
gas and air filtration	气体和空气过滤
deceleration of vehicle	车辆减速
acoustic and vibration control	声学与振动控制
thermal and acoustic protection	放热防噪音保护
flame resistant characteristics	阻燃特性
pressure vessel	压力容器
survivable limits	可存活极限
load carrying capacity	承载能力
uniform rubber mass	均匀橡胶质量
dimensional stability	尺寸稳定性

Critical reading and thinking

Task 1 Overview

Work in pairs and retell the function of medical textiles. Use as many lexical

chunks as possible.

Task 2 Group discussion

Work in groups of 4-5 and have a discussion about the following questions.

1. Can you imagine what functions technical textiles play in aerospace?
2. What problems do you think the production of technical textiles has in China?

Task 3 Language building-up

I. **Translate the following terms from English into Chinese or vice versa.**

artificial kidney	
durable ballistic fabric	
implantable medical textiles	
ultra-violet ray	
glass fibre	
人工肺	
安全带	
低温(织物)	
医用织物	
生物危害防护服	

II. **Complete the following Sentences with the words or phrases given in the previous exercise.**

1. _____ is made with hollow polypropylene fibre (聚丙烯纤维) or a hollow silicone membrane(中空硅树脂膜) and used to remove carbon dioxide from patient's blood and supply fresh oxygen.
2. The modern plaster fabric is made from spun bonded nonwovens of cotton (纺粘非织造布), viscose, polyester or _____.
3. Studies have carried out to impart the properties of wound care and antibacterial finishes in _____.
4. _____ has been one of the leading high temperature thermal insulation materials.
5. _____ is an energy absorbing device, which controls the forward movement of wearer in an event of sudden deceleration of vehicle.

Task 4 Guessing game

Guess the missing lexical chunks from different contexts with the ones in the box.

> A. glass fibre   B. plastic surgery   C. insulation materials

Group 1
1. As with any surgery, _____ requires conditions of optimum safety and experienced practitioners.
2. She discusses how _____ not only keeps her looking youthful, but feminine as well.
3. Device which cool the skin aims to relieve pain caused by various medical laser treatment, injuries, _____ and minor surgery.

Group 2
1. _____ run the gamut from bulky fibre materials such as fibreglass, rock and slag wool, cellulose, and natural fibres to rigid foam boards to sleek foils.
2. The most important part about picking a hot _____ is understanding the maximum temperature the insulation will be covering.
3. Other _____ not mentioned are natural fibres such as hemp, sheep's wool, cotton and straw.

Group 3
1. Typically the _____ have diametre between 3.8 and 20μm.
2. The leading types of _____ are E-glass, high-strength (HS)-glass, and corrosion resistant (CR)-glass.
3. _____ is made by melting the constituent materials together and drawing the melt into a fibre.

Task 5 Translation

Translate the following paragraphs into Chinese.

   In the 21st century, we have witnessed the proliferation of connected cars, homes, and countless other devices. Smart home devices allow for remote programing, remote function control, and remote diagnostic of the home's various traditional systems. Lighting, home security, heating, and A/C are all controlled through home automation hubs, smart phones, or voice assistants. Given the growing connection of so many everyday devices, it was only a matter of time until it reached our clothing: Specifically, the creation of garments that are rich in data, provide user feedback, and can be connected to other digital devices.

   Advances in micro-sensor technology and textiles are setting the stage for further breakthrough in how we not only wear but also engage with our clothes. Innovative breakthrough in the textile industry has traditionally occurred slowly. For tens of thousands of years, humanity has relied primarily on animal pelts and plant fibres like linen, cotton, leather, silk, and hemp as materials for clothing.

Task 6 Research

Surf the Internet for more information about the specific application of technical textiles.

(You may refer to textilelearner.net for more information with your cell phone or you can visit http://www.textilesintelligence.com for more about technical textiles intelligence.)

Section 3　Smart Fabrics

Lead-in

Warm-up questions:

Now we have 10 billion smart devices connecting and exchanging data with other devices and systems over the Internet. Some cherish this Internet of Things (IoT), and others are quite anxious. What's your opinion?

Lexical chunk bank	
力量型西装	power suit
电子纺织品	electronic textile
(新技术的)早期使用者；早期接受者	early adopter
初创公司，新成立的公司	startup
家用电器	household appliance
周边产品；搭车产品	spin-off
确定的订货；定期订货	firm order
导电	conduct electricity
导电的；导电性能好	electrically conductive
电磁屏蔽	electromagnetic shielding
高分子电子纺织材料	textro-polymer
太阳能电池板	solar panel
为了；着眼于	with a view to
控制线路；控制电路；控制回路	control circuit
时尚设计酒店；设计师酒店	designer hotel
电压梯度；电压陡度	voltage gradient
湿敏元件；湿度传感器	moisture sensor

大小便失禁	incontinence
输出设备；输出装置	output device
发光二极管	light-emitting diode（LED）
热敏油墨	thermochromic ink
通上电流；加电	applying a current
折中，妥协	halfway house
电干扰；电气干扰	electrical interference
电源；供电	power supplies
融合	blur the lines
浑然一体	blend into the background

It could give the term **power suit** a whole new meaning. Getting dressed a few years from now, you may find yourself putting on more than mere fabric. Your clothes may by then **sport** electronic **sensors** and tiny computers. As you walk out of the door, you will be not just fashionably **attired**, but digitally enhanced a living, breathing **node** on the Internet. This prospect will delight some people and horrify others. But it could actually happen, if the field known variously as smart fabrics, **electronic textiles** or washable computing can achieve the breakthrough its **proponents** believe is just around the corner.

As recently as five years ago the idea of clothing, furniture and upholstery that combined fabric with electronics was a fantasy. Yet today the first examples of the technology are on sale, with more advanced products on the way. Current products are aimed at **early adopters**, but both hopeful **startups** and big firms such as Nike, DuPont and Philips are searching for an application that will carry the technology into the mainstream.

Smart fabrics look and feel like ordinary textiles, but can do extraordinary things: generate heat, **monitor** vital signs, act as switches or sensors, and even change color. With so much fabric woven into daily life, proponents of smart fabrics see them as a natural way to increase the **pervasiveness** of today's **gadgets** and add **snippets** of intelligence to everyday items. Computing power is already being incorporated into cars, **household appliances** and entertainment systems, notes Stacey Burr, the boss of Textronics©, a **spin-off** from DuPont based in Wilmington, Delaware, that is developing electronic textiles and clothing. So fabric, she argues, is a natural next step.

About 70% of the materials that people come in contact with are fabrics, she says. We want to create fabrics that warm, **illuminate**, conduct, sense and respond. Smart fabrics will be particularly useful in the fields of medicine, sports, communications and personal security, Ms Burr predicts.

She is not alone in her enthusiasm. There is a really big market, says John Collins of Eleksen, a British startup that sells sensors based on smart fabrics that can be incorporated into clothing, accessories and furniture. Eleksen has so far sold 70,000 units, but already has **firm orders** for 600,000 units this year, and expects sales to rise tenfold again soon.

Smart fabrics can take many forms. The most basic kind is **electrically conductive**, such as Textronics' *textro-yarn*, a slightly **elastic** material that **resembles** ordinary fabric. Because it **conducts electricity**, it can be used for heating (by passing a **current** through the fabric), as a radio **antenna**, for **electromagnetic shielding**, to provide power to other devices **embedded** into clothing, and even to make electrodes, for example to monitor vital signs. The company has just launched a sports bra that monitors the wearer's heart-rate and calorie consumption, and displays them on a wristwatch-sized screen.

Another of Textronics' smart materials is **textro-polymer**. Its fibres have the useful property that their **resistance** changes when they are stretched. This can be used to detect bending, stretching or tugging, which can in turn reveal whether the wearer of a smart garment is moving or **stationary**, or whether a particular car seat or bed is **occupied**. In September, Textronics announced a deal with Konarka, a pioneer in flexible **solar panels**, **with a view to** making jackets that can recharge mobile phones and other devices.

International Fashion Machines (IFM), a firm based in Seattle, has just launched a range of light switches based on conductive yarns. Squeezing the fuzzy, **pompom**-shaped blob of material changes the amount of current flowing through it. This difference is detected by a **control circuit** that then turns lights on or off accordingly. Maggie Orth, IFM's founder, hopes these playful switches will find use in children's playrooms and **designer hotels**.

The smart fabrics made by Eleksen, called ElekTex, acts as a more elaborate touch sensor. It consists of three layers: a top and bottom layer of conductive material, and a middle layer that conducts electricity when it is compressed. **Voltage gradients** are applied across the top and bottom layers, at right angles to each other. When the fabric is pressed, current flows through the middle layer. By measuring the change in voltage across the top and bottom layers, it is possible to determine where (and roughly how hard) the fabric is being pressed. This means that sliding and tracing gestures, as well as individual presses, can be detected.

Eleksen claims that the fabrics can survive being washed, **crumpled, punctured** and even driven over. It has already been used to make roll-up fabric keyboards for handheld computers, to control a heated jacket made by Innovative Sports, and to incorporate iPod controls into ski jackets made by Spyder, Kenpo and Westcomb. The iPod is popped into a special pocket and plugged into a control wire, and it can then be controlled using fabric sensors in the jacket's cuff or on its sleeve. Buttons start and stop playback and select tracks, and **stroking** a strip adjusts the volume.

In France, a cinema has experimented with using ElekTex to count the number of occupied seats. Another proposed use is to incorporate television controls directly into the fabric of sofas. The real value in our stuff is being able to keep the **interface** as soft as the thing it's going into, says Miles Jordan of Eleksen. The company has also devised a second form of smart fabric, again made up of several layers, which functions as a **moisture sensor**. The top two layers allow only a small proportion of **incident** moisture to reach the third, bottom layer, which contains a matrix of conductive fibres. The amount of moisture can be determined by measuring variations in the resistance of this layer. This smart fabric can be used for what Eleksen delicately calls **incontinence** detection in medicine, and to detect moisture in buildings.

As well as acting as sensors and switches that gather information, smart fabrics can also function as **output devices**. One approach is that taken by Luminex, the result of a collaboration between Caen, an Italian electronics firm, and Stabio, a Swiss textiles company. Luminex is a fabric with fibre-optic strands woven into it, which are then illuminated using **light-emitting diodes** powered by a small battery pack. Luminex has already been incorporated into glowing clothes, safety garments, handbags and furniture, and even a wedding dress.

Electric Plaid, devised by IFM, takes a different approach. Rather than emitting light directly, it contains **stainless-steel** yarns coated with **thermochromic inks** that, as their name suggests, change color depending on the temperature. **Applying a current** causes the yarns to heat up, which changes the ink's color. This makes possible fabrics with slowly changing patterns, and even information displays: a **wall-hanging** that changes color depending on the weather forecast, for example.

As smart as these fabrics are, they still rely on separate control circuitry to detect pressure, motion or moisture, or change their appearance. Smart jackets based on ElekTex, for example, contain a small control unit to connect the touch sensors on the sleeve to an iPod. Rehmi Post, a researcher at the Centre for Bits and Atoms at the Massachusetts Institute of Technology (MIT), says the real breakthrough will come when the control electronics are not simply housed in clothing, but are woven directly into the fabric. Today's products, he points out, are a **halfway house** in which the textiles and electronics are not fully **integrated**. They aren't ambitious enough, he says.

Seamlessly weaving cloth, power and data, Dr Post admits, will be easier said than done. Many of the required technologies exist already, but overcoming problems such as **electrical interference**, programming and **power supplies** will not be trivial. There are a few fundamental problems that have to be solved, but they are solvable, he says.

Integrating fabrics and electronics more closely will, Mr Collins predicts, make possible phones, music players and other portable devices that are part electronic and part fabric. They will, he says, be smaller, lighter, less power-hungry and more durable than today's devices.

Smart clothing could not only **blur the lines** between materials and electronics but, if items of clothing start to absorb previously **discrete** devices between people and machines, they will be a kind of second skin and functionality will **blend into the background**, says Ms Burr. Perhaps. But in the short term, at least, it seems more likely that smart clothing's appeal will be limited to particular, well-defined situations, such as skiing, policing and emergency rescue. That said, many people now refuse to go out without their mobile phones. The challenge for believers in smart fabrics is to make people feel similarly naked without them.

Words：

sport [spɔː(r)t]	v. 夸示；得意地穿戴；故意显示
sensor ['sensə(r)]	n.（探测光、热、压力等的）传感器
attired [ə'taɪə(r)d]	adj.〈正式〉身着；穿戴着
node [nəʊd]	n.（网络）节点；交点；（根或枝上的）节
proponent [prə'pəʊnənt]	n. 支持者；倡导者；拥护者
monitor ['mɒnɪtə]	v. 监控；监视；监听；检查
pervasiveness [pə'veɪsɪvnəs]	n. 无处不在，遍布
gadget ['gædʒɪt]	n. 小玩意儿；小配件；新科技
snippet ['snɪpɪt]	n.（消息、知识等）的片断；零星的活
illuminate [ɪ'luːmɪneɪt]	v. 照亮；照明
elastic [ɪ'læstɪk]	adj. 有弹性的；有弹力的
resemble [rɪ'zembl]	v. 类似；相似；相像
current ['kʌrənt]	n. 电流；水流；气流
antenna [æn'tenə]	n. 天线；接收天线；[生] 触角；触须
embed [ɪm'bed]	v. 嵌入；内嵌；埋置
resistance [rɪ'zɪst(ə)ns]	n. 电阻；抵抗；阻力；抗力
stationary ['steɪʃ(ə)n(ə)ri]	adj. 不动的；静止的；固定的
occupied ['ɒkjʊpaɪd]	adj. 有人使用（或居住）；使用中
pompom ['pɒmpɒm]	n. 小绒球；（美国啦啦队常用的）塑料丝球
crumple ['krʌmpld]	v. 揉皱；变皱

puncture [ˈpʌŋktʃə(r)]	v. 击穿；穿孔；(车胎等的)刺孔
stroke [strəʊk]	v. 抚摸；轻触；[纺] 弄伸皱褶
interface [ˈɪntə(r)feɪs]	n. 接口；连接电路
incident [ˈɪnsɪd(ə)nt]	adj. 附带的；难免发生的，伴随的
stainless-steel [ˌsteɪnləs ˈstiːl]	n. 不锈钢
wall-hanging [wɔːlˈhæŋɪŋ]	n. 墙帷；挂饰
integrated [ˈɪntɪɡreɪtɪd]	adj. 集成的；各部分密切协调的；综合的
discrete [dɪˈskriːt]	adj. 离散的；分离的；不相关联的

Critical reading and thinking

Task 1 Overview

Work in pairs and discuss the development and application of smart fabrics. Use as many lexical chunks as possible.

Task 2 Group discussion

Work in groups of 4-5 and have a discussion about the following questions.

1. What kind of new applications will you propose for the smart fabrics? How to realize them?
2. What is the bottleneck in the development of smart fabrics?

Task 3 Language building-up

Translate the following terms from English into Chinese or vice versa.

startup	
spin-off	
electrically conductive	
applying a current	
power supplies	
strand	

电子纺织品	
湿敏元件	
热敏油墨	
电阻	

Task 4　Research

Surf the Internet for more information about the smart fabrics.
(You can try https：//www. sohu. com/a/455170790_116132, https：//www. sciencedirect. com/topics/engineering/electronic-textile.)

Glossary

a myriad of	大量的	C2S2
Abba Gould Woolson	乌尔森（美国的女性存在主义教母）	C3S1
abrasion	摩擦	C4S3
abrasive substance	研磨材料	C3S1
absorbency	吸收性	C4S1
absorbent	能吸收的；有吸收力的；吸收剂；中和剂	C6S2
absorbing layer	吸收层	C6S2
accessory	配件，配饰	C2S1
accord	给予，赠予，授予（权力、地位、某种待遇）	C1S2
accoutrements	〈正式〉装备；配备	C1S1
acetate	醋酸纤维素及其制成的产品	C3S2
acetate jersey	醋酸酯平纹针织衫	C3S2
acid	酸；酸性物质	C6S1
acoustic and vibration control	声学与振动控制	C6S2
acrylic	腈纶；聚丙烯腈	C4S1
acute	敏感的；敏锐的；剧烈的	C6S2
adage	格言	C2S1
adapt	使适应；改变；适应	C5S2
additive	（尤指食品的）添加剂，添加物	C5S3
adhesive	黏合剂，黏着剂	C4S5
adhesive film	胶膜，黏合膜	C4S5
adhesive material	粘合材料	C4S5

adjacent	相邻的	C4S4
admission	招生	C1S2
advanced EOD bomb suit	高级爆炸物处理炸弹套装	C6S2
aesthetic	美学的；审美的；有美感的	C6S1
air jet loom	喷气织机	C4S3
align	对准，校直	C4S2
alpaca	（南美的）羊驼，羊驼毛，羊驼呢(织物)	C4S1
alumnus	毕业生；校友	C1S2
ambition	追求的目标；夙愿；抱负，雄心	C5S3
animal fibre	动物纤维	C4S1
anomaly	反常之事物；不规则	C6S1
antenna	天线；接收天线；[生]触角；触须	C6S3
antibacterial	抗菌的	C4S1
antibacterial finishes	抗菌整理	C6S2
apparel	服装；衣服	C3S1
apparel system	服饰制度	C2S2
apparel textiles	服装纺织品	C6S1
applying a current	通上电流；加电	C6S3
apron	围裙	C2S2
archaeological	考古学的	C2S1
archeological	考古学的	C4S4
archeologist	考古学家	C2S1
archives	档案；档案馆(室)	C1S2
array	一系列；大批；数组；陈列	C6S1
artefact	文物；人工制品，手工艺品(尤指有历史或文化价值的)	C1S1
artifact	人工制品，手工艺品	C3S1
artificial intelligence	人工智能	C5S2

artificial (kidney, liver, heart)	人工(肾脏、肝脏、心脏)	C6S2
artistic textiles	艺术纺织品	C2S1
asbestos	石棉	C4S1
Asia Minor	小亚细亚	C5S1
at par (with)	与票面价值相等；媲美	C1S2
atlas stitch	经缎组织	C4S4
attire	服装；盛装	C3S1
attired	〈正式〉身着；穿戴着	C6S3
auditorium	大会堂；礼堂；〈美〉讲堂	C1S2
auspicious	吉利的；吉祥的	C1S1
aviation	航空；飞机制造业	C6S1
ballistic	弹道(学)的；飞行物体的	C6S2
bandage	绷带	C6S1
barbed needle	刺针，倒钩针	C4S5
barrier materials	阻隔材料	C6S1
bast fibre	韧皮纤维；麻纤维	C1S1
batik	蜡染	C2S1
BCE: before the Christ Era	公元前	C5S1
beard	(钩针)针钩	C4S4
bearded/spring needle	钩针/弹簧针	C4S4
beating up/in	打纬	C4S3
bequest	遗产；遗赠	C1S2
bestow	赐；授予；给予	C1S2
Bian Embroidery	汴绣	C2S2
biohazard suit	生物危害防护服	C6S2
biological organism	生物有机体	C4S1

bleach	漂白剂；漂白；变白；漂白	C6S2
blend into the background	浑然一体	C6S3
block printing	手工木板印花	C2S1
blockchain technology	区块链技术	C5S2
blow room	吹风室，滤尘室	C4S2
blue and white gingham	蓝白方格布	C3S2
blunt impact	钝性冲击	C6S2
blur the lines b/n	融合	C6S3
bobbin	线轴，绕线筒	C4S2
bordered	镶边的；有装饰边；织花边的	C1S1
boycott	联合抵制；拒绝购买；拒绝参加	C5S1
breathability	透气性	C4S1
breathable fabric	透气织物	C3S2
breech	臀部；后膛	C3S1
bristle	鬃毛；刚毛	C4S2
brocade	织锦缎；（尤指用金银线织出凸纹的）厚织物	C1S1
brooch	（女用的）胸针，领针	C3S1
bruising	挫伤	C6S1
bulletproof vest	防弹背心	C6S1
bunched heald jacquard	束综提花机	C2S2
bureaucrat	官僚	C2S1
bursting strength	爆裂强度	C6S1
bustling	熙熙攘攘的，忙乱的	C2S1
butt	针踵	C4S4
Byzantines	拜占庭人	C3S1
calender	用砑光机压光	C2S1
car interior	汽车内饰	C6S1
carbon fibre	碳纤维	C4S1

carbonization	碳化	C4S1
card frame	梳棉机	C4S2
card sliver	粗梳生条	C4S2
carded yarn	精梳纱	C4S2
cardigan	开襟绒线衫；羊毛背心；羊毛衫	C4S4
carding	梳理(棉、毛、麻等)；梳毛	C4S2
cashmere	开司米；山羊绒；克什米尔羊毛	C3S2
cashmere fibres	羊绒纤维	C3S2
catalyze	催化；刺激，促进	C5S2
catastrophic	灾难的；悲惨的；灾难性的，毁灭性的	C5S3
cavalry	骑兵	C5S1
cellulose	纤维素	C4S1
cellulose fibre	纤维素纤维	C4S1
ceramic fibre	陶瓷纤维	C4S1
chambray	有条纹或格子花纹的布	C3S2
chapel	(学校、监狱、私人宅院等基督教徒礼拜用的)小教堂	C1S2
chart	用图表示(说明)	C1S2
chemical binder	化学黏合剂	C4S5
chemical bonding	化学黏合	C4S5
chemical solvent	化学溶剂	C4S1
chemical warfare agents	化学战剂	C6S2
chiffon	薄绸；雪纺绸	C3S2
chiton	希顿古装(古希腊人贴身穿的宽大长袍)	C3S1
chitosan fibre	甲壳素纤维	C4S1
chlamys	古希腊男子所着的一种短斗篷或外套	C3S1
cinnamon	肉桂	C5S1
civil engineering	土木工程	C6S1

classical antiquity	古典时代	C2S1
clearing	退圈	C4S4
clinch	确定，敲定，解决	C3S2
closed lap	闭口式垫纱，闭口线圈	C4S4
closing	闭口	C4S4
Cloud Brocade	云锦	C2S2
Coco Chanel	可可·香奈儿	C4S4
cocoon	蚕茧	C2S1
cohesiveness	抱合力；黏结性，内聚性	C4S1
coin	创造（新词语）	C5S1
collision	猛烈相撞；抵触；碰撞；互撞	C6S2
colonisation	殖民地化	C4S4
colour symbolism	色彩象征	C3S2
combed yarn	普梳纱	C4S2
combination weaves	复合机织组织	C4S3
combing	梳毛，梳理	C4S2
commercial value	商业价值	C2S1
compatibility	和谐共处；协调；兼容	C6S1
complementary	补足（充）的	C6S2
complex	（类型相似的）建筑群，综合体，大楼	C1S2
composite	合成物；复合材料	C6S1
compound needle	复合针	C4S4
compress	压缩；压榨；镇压	C6S2
compression bandage	压缩绷带	C6S2
compression garment	压缩服	C6S1
constitute	构成	C5S2
contamination	污染；污秽；污物；混淆不清；污秽物	C6S2
control circuit	控制线路；控制电路；控制回路	C6S3

converge	使汇聚；聚集；靠拢；收敛	C5S3
convergent	会合的；逐渐减小的	C4S5
cool colours	冷色	C3S2
cord	绳	C4S1
cost-effective	成本效益	C5S2
cotton blend fabric	棉混纺织物	C3S2
cotton blends	棉混纺	C3S2
cotton roving	面粗纱	C2S1
cotton textile industry	棉纺织业	C2S1
course	横列	C4S4
COVID-19	新冠肺炎（2019冠状病毒病）	C5S2
crack	破裂；砸	C6S1
craft of designing or creating textiles	纺织品设计、制造工艺	C2S1
credit	信用，赞许，学分；(~someone with sth.)把(某成果)归因于某人	C5S1
crepe	绉纱；绉绸	C3S2
crepe yarns	绉纱	C3S2
crimp	卷曲	C4S1
crocheting	钩针编织	C1S1
cross-laid	交叉放置的	C4S5
cross-sectional	截面的，断面的，剖面的	C4S1
crumple	揉皱；变皱	C6S3
cuff	袖口	C4S4
cultivated silk	家蚕丝；桑蚕丝	C4S1
cultural heritage	文化遗产	C2S2
cultural transformation	文化转型	C2S2
curl or roll	卷边	C4S4

current	电流；水流；气流	C6S3
dacron	涤纶；的确良	C4S1
dalmatica	教士法衣	C3S1
damask	锦缎；花缎	C1S1
dark colors contract	深色收缩	C3S2
debris	残骸，碎片；垃圾	C4S2
deceleration of vehicle	车辆减速	C6S2
decitex (dtex)	分特克斯	C4S2
decree	裁定；判决；颁布(法令)	C1S2
defect	缺陷；过失；缺点；弱点，瑕疵	C6S2
degum	脱胶；水洗	C1S1
dcnier	旦尼尔	C4S2
denim	斜纹粗棉布，牛仔布	C4S3
derivative weaves	变化机织组织	C4S3
designate	命名，指定	C4S2
designer hotel	时尚设计酒店；设计师酒店	C6S3
diagonal	斜纹的；对角线的；斜线的	C4S3
die	模具	C4S5
dimensional	维度的；次元的；空间的；尺寸的	C6S2
dimensional stability	尺寸稳定性	C6S2
dimity	凸花条纹布	C3S2
discrete	离散的；分离的；不相关联的	C6S3
disposable income	可支配收入	C5S2
disposable medical/hygiene product	一次性医疗/卫生产品	C4S5
disruption	扰乱，打乱，中断	C5S3
domestic water sewerage plants	生活污水处理厂	C6S1

219

double faced fabric	双面织物	C4S4
drainage	排水；排水系统；污水	C6S1
drainage system	排水系统	C6S1
drape	用布帘覆盖；使呈褶裥状；悬垂	C3S1
drapery	布料；帏帐；打褶的帐幔	C3S2
draw frame	并条机	C4S2
drawboy	挽花工	C1S1
drawing-in	穿经	C4S3
drawloom	手工提花织机；花机	C1S1
drawn sliver	熟条	C4S2
dressing	敷料	C4S2
dropper	停经片	C4S3
dry spinning	干法纺丝	C4S2
durability	持久性；耐久性	C6S1
durable ballistic fabric	耐久弹道织物	C6S2
dyed yarn woven	机织染色纱	C3S2
early adopter	(新技术的)早期使用者；早期接受者	C6S3
ecological protection	生态保护	C6S1
ecological reform	生态改革	C6S2
economic staple	经济支柱	C2S1
Egyptian cotton	埃及棉	C4S1
elastic	有弹性的；有弹力的	C6S3
elastic recovery	回弹性	C4S1
electrical interference	电干扰；电气干扰	C6S3
electrically conductive	导电的；导电性能好	C6S3
electromagnetic shielding	电磁屏蔽	C6S3
electronic textile	电子纺织品	C6S3
electrostatic charge	静电荷	C4S5

elements	(尤指恶劣的)天气	C4S4
elongation	伸长；伸长率	C4S1
embankment	堤防；路基；堤岸；(铁路的)路堤	C6S1
embed	嵌入；内嵌；埋置	C6S3
embellish	装饰；修饰	C2S1
embellishment	装饰品	C2S1
embroider	刺绣；绣花	C1S1
embroidered design	刺绣图案	C1S1
embroidery	刺绣	C2S1
emerging	新兴的；脱颖而出的；日益壮大的	C1S2
emissary	使者；密使	C5S1
emulsion	乳状液	C4S5
en bloc	整体；全部；一起；统统	C1S2
engine sound isolation	发动机隔音	C6S1
engineered fabric	工程织物	C4S5
entangle	(使)纠缠，缠住	C4S5
erosion	腐蚀；侵蚀；流失	C6S1
ethnic cultures	民族文化	C2S2
evenness	平整度；均匀度	C4S2
evocative	引起记忆的；唤起感情的	C1S1
exemplary	典范的；可作榜样的；可作楷模的	C1S2
existing color palette	现有调色板	C3S2
exotic	奇异的	C5S1
expedition	远征；探险队	C5S1
exquisite	精致的	C2S2
extant	尚存的；现存的；未遭毁灭的	C1S1
extensibility	延伸性	C4S4
externality	外部效应	C5S3

extra corporeal device	体外装置	C6S2
extract	提取，取出	C2S1
eye	针眼	C4S4
eye cut	针舌槽	C4S4
fabric	织物；布料	C1S1
fabric defect	织疵	C4S3
fabric imperious	织物专制力	C6S2
facade foundation	立面基础	C6S1
face loop stitch	正面线圈组织	C4S4
facilitate	（使）容易；（使）便利；促进；帮助	C6S2
faculty	全体教员；学院；系	C1S2
fancy woven stripe	花式编织条纹	C3S2
fashion and apparel design	时装设计专业	C1S2
fashion and textile management	市场与纺织品管理专业	C1S2
fashion color predictions	流行色预测	C3S2
fashion colours	时尚色彩	C3S2
feeding	垫纱	C4S4
feedstock	（制造产品的）原料	C5S3
fell	织口	C4S3
Ferdinand von Richthofen	斐迪南·冯·李希霍芬	C5S1
fibre composites	纤维复合材料	C6S2
fibroin	蚕丝蛋白	C4S1
fibrous	含纤维的，纤维性的	C4S1
filament	长丝	C4S1
filament fibre	长丝纤维	C4S1
filament yarn/thrown yarn	长丝纱	C4S2
filter	过滤；渗透	C4S2

filter clothing	滤布	C6S1
filtration	过滤；滤清	C6S2
fineness	细度	C4S2
finish	整理	C4S1
finishes	成品（finish 的复数）	C3S2
firm order	确定的订货；定期订货	C6S3
flame resistant characteristics	阻燃特性	C6S2
flame retardant	阻燃的	C4S1
flammable	易燃的，可燃的	C4S1
flat heald jacquard	水平织机，平综提花机	C2S2
flax	亚麻；亚麻纤维	C1S1
flax plant	亚麻树	C3S2
fleece	羊毛，绒头织物	C3S2
flexibility	柔韧性	C4S1
flock	很短的纤维	C4S2
foliage	（植物的）叶；枝叶	C1S1
fossil fuel	化石燃料	C5S3
foster	促进；助长；培养；鼓励	C5S2
foul	污浊的；恶臭的；险恶的；（言辞）粗俗的	C6S2
front protection plate	前护板	C6S2
frontal	额前装饰物（如发带、头帕）；（祭坛前面的）帷子	C1S1
fuel	给……提供燃料；刺激，煽动；推动	C5S2
fuel economy	燃油经济性	C6S2
full dress	全套连衣裙	C3S1
functionality	设计目的；设计功能，实用	C5S3
fur garments	毛皮服装	C3S1
furnishings	家具陈设	C1S1

fuselage	机身	C2S2
gabardine	一种斜纹防水布料，华达呢	C4S3
gadget	小玩意儿；小配件；新科技	C6S3
garment	衣服，服装	C3S1
gas and air filtration	气体和空气过滤	C6S2
gauge	机号	C4S4
gauze	薄纱，纱罗织物	C1S1
gauze weave	纱罗织物	C1S1
General Assembly	（美国的）州议会	C1S2
genetically modified cotton (GM)	转基因棉	C4S1
genus	（动植物的）属；类；种；型	C4S1
geotextile	土工织物，土工布	C4S5
ginger	姜	C5S1
gingham	条纹棉布；条格平布	C3S2
glass fibre	玻璃纤维	C4S1
Godey's *Lady's Book*	戈迪的《女书》（19世纪美国的一本杂志）	C3S1
goggle	眼睛睁视；护目镜	C6S1
governor	（学校、学院、医院等机构的）董事，理事；负责人	C1S2
gown	女长服；长外衣；外罩	C1S1
grant	允许；同意；（政府、机构的）拨款	C1S2
grinding technology	研磨技术	C6S1
ground material, ground fabric	底布	C1S1
Gu Embroidery	顾绣	C2S2
guarantee	保证（人）；保证书；保证；承诺；确定；保护	C6S2
guide bar	导纱梳栉	C4S4
gum	黏胶，胶质物（用以粘轻东西，如纸等）	C1S1

hail	冰雹	C6S1
halfway house	折中，妥协	C6S3
Han Dynasty of China	汉朝	C5S1
Han Embroidery	汉绣	C2S2
harass	骚扰，侵扰，不断攻击	C5S1
harness	综框	C4S3
hazardous	有危害的	C5S3
head	针头	C4S4
headgear	帽子	C2S1
headliner carpets	车顶衬里地毯	C6S2
heald	综；综框；综线	C2S2
heat retention property	保暖性	C4S4
heated roll or cylinder	热滚筒	C4S5
heddle	综丝；综片综线	C4S3
heddle eye	综眼	C4S3
helical	螺旋状的	C4S2
hemp	大麻	C1S1
high added-value	高附加值	C5S2
high performance textile materials	高性能纺织材料	C6S2
high tenacity	高韧性	C6S1
high-waisted dresses	高腰连衣裙	C3S1
hollow viscose	中空粘胶	C6S2
hook	针钩	C4S4
horizontal loom, ground loom	卧式织机；踞织机	C1S1
hosiery	袜类；针织内衣	C4S4
hostel	〈美〉(招待徒步旅行青年等的)招待所；〈英〉大学宿舍	C1S2

household appliance	家用电器	C6S3
hub	(某地或活动的)中心；枢纽	C1S2
hub and spoke structure	中心辐射结构	C1S2
human organic tissue	人体器官组织	C6S1
hydroentangled nonwoven fabric	水刺无纺布	C4S5
hydroentanglement	水刺(射流缠结)	C4S5
hygiene	卫生；卫生学	C6S1
hygiene medicals	保健药品	C6S1
hygienic	卫生的，保健的；卫生学的	C3S1
illuminate	照亮；照明	C6S3
impact protection fabrics	防冲击织物	C6S2
impart	传授；告诉	C1S2
imperative	重要紧急的事；必要的事	C2S1
impermeable materials	防渗材料	C6S2
implantable medical textiles	植入性医用纺织品	C6S2
impose	加(负担、惩罚等)于；课(税)；处(罚)	C6S2
impurity	杂质	C4S2
inadvertently	无意地；不经意地	C5S3
inaugural	就职的；开幕的；成立的；创始的	C1S2
incident	附带的；难免发生的，伴随的	C6S3
incinerate	焚毁，把……烧成灰烬	C5S3
incontinence	大小便失禁	C6S3
incorporate	合并	C6S1
indigo	靛蓝；靛青	C2S1
industrial textiles	工业纺织品	C6S1
industry-wide	全行业的	C5S3
inexorably	不可逆转地，不可阻挡地	C5S3

initiative	倡议；新方案	C1S2
inorganic fibre	无机纤维	C4S1
instantaneous	立刻的，瞬间的；即时的	C6S2
instrumental	乐器的；有帮助的	C6S1
insulate	(使)绝缘，(使)隔热	C4S2
insulation materials	绝缘材料	C6S2
insulator	绝缘、隔热或隔音等的物质或装置	C4S1
insulin	胰岛素	C1S2
integrated	集成的；各部分密切协调的；综合的	C6S3
interface	接口；连接电路	C6S3
interlace	(使)交错，(使)交织	C4S2
interlock fabric	双罗纹织物	C4S4
interloop	圈套在一起	C4S4
intermediary	中间的；媒介的；中途的；中间人；仲裁者；调解者；媒介物	C5S1
intermesh	(使)互相结合，(使)互相啮合	C4S3
intersection	横断，横切	C4S5
interspersed	点缀的	C2S2
intertwine	缠结在一起；使缠结	C4S5
isolation gowns	隔离衣	C6S1
jersey	运动衫，毛线衫	C3S2
jersey knit fabric	平纹针织物	C3S2
Kashmir	克什米尔	C4S1
khaki	卡其布	C4S3
kick-start	促使……开始；使(项目)尽快启动	C5S3
knitted technology	针织工艺	C4S5
knitted textile	针织纺织品	C4S1
knitwear	针织品	C4S4

knocking-over	脱圈	C4S4
Kublai Khan	元世祖（忽必烈汗）	C5S1
labour-intensive sector	劳动密集型行业	C5S2
laminate	层压材料；叠层，层压	C4S5
landfill	废物填埋地	C5S3
land-grant college	（美国）政府资助的低学费大学	C1S2
landing	套圈	C4S4
landscape gardening	园林绿化	C6S1
lap	垫纱	C4S4
lashing straps	系带	C6S1
latch	针舌；门闩，插销	C4S4
latch needle	舌针	C4S4
lateral	横向的；侧面的	C4S3
lead time	交货期	C5S2
leakage	漏；泄漏	C6S2
leftover	剩饭；残留物	C6S1
leg wrappers	裹腿裤	C3S1
Leizu	嫘祖	C2S1
length based system/fixed weight system	定重制	C4S2
length-to-width ratio	长径比	C4S1
letting off	送经	C4S3
Li Brocade	黎锦	C2S2
light cotton clothing	轻薄棉衣	C3S1
light needle punching	轻型针刺	C6S2
light support bandage	轻型支撑绷带	C6S2
light-emitting diode (LED)	发光两极管	C6S3
lightweight fabric	轻质织物	C3S2

linear	线的；直线的；线状的	C5S3
linear density	线密度	C4S2
linen	亚麻布	C2S2
linen cloth	亚麻布	C3S1
linen fabric	亚麻织物	C3S2
lining	内衬；衬里；内胆	C2S1
literary temperament	文学气质	C2S2
livestock	家畜；牲畜	C6S1
load carrying capacity	承载能力	C6S2
locomotive	机车；火车头	C1S2
loft	弹性；蓬松	C4S1
logistics	物流	C5S3
longitudinal	经度的；纵向的；纵的；纵观的	C4S1
loop	线圈；(使)成圈；以环联结	C4S4
loop formation	成圈	C4S4
lubricant	润滑剂，润滑油	C4S2
lustre	光泽	C2S1
lustrous	柔软光亮的	C1S1
mandatory	法定的；义务的；强制性的；受托管理者	C6S2
Marco Polo	马可·波罗	C5S1
mass	团，质量，大量	C6S2
maximum	最大量；最大值；最高(大)的；最大值的	C6S2
mechanical engineering	机械工程	C6S1
mechanical organ	机械器官	C6S2
medical textiles	医用纺织品	C6S1
medication	药疗法；加入药物；药物处理；医药，药物	C6S2
medium of exchange	交换媒介	C2S1
melt spinning	熔体纺丝	C4S2

meltblown nonwoven fabric	熔喷无纺布	C4S5
meltblown technology	熔喷工艺	C4S5
merge	合并；融合；归并	C1S2
Mesopotamia	美索不达米亚	C5S1
metal brooches	金属胸针	C3S1
metal fibre	金属纤维	C4S1
micrometre	微米	C4S1
microorganism	微生物	C4S1
microtherm	低温（织物）	C6S2
mildew	霉	C4S1
milk fibre	牛奶纤维	C4S1
millennia	千年期（millennium 复数）	C2S1
millennium	千年期	C2S1
mineral fibre	矿物纤维	C4S1
Modal	莫代尔纤维	C4S1
moisture absorbency	吸湿性	C4S1
moisture absorption	吸湿	C6S1
moisture regain	回潮率	C4S1
moisture sensor	湿敏元件；湿度传感器	C6S3
molting	换羽，脱毛；蜕皮	C4S1
Mongolian Empire	元朝（蒙古帝国）	C5S1
monitor	监控；监视；监听；检查	C6S3
mosque	清真寺	C1S2
motif	装饰图案；装饰图形	C1S1
mount calligraphy and painting	书画装裱	C2S2
mulberry silk	桑蚕丝	C4S1

multilateral development banks	多边开发银行	C5S2
multi-spool machine	多轴纺纱机	C2S1
natural fibres	天然纤维	C3S1
navy blue	藏青蓝	C3S2
necessitate	使……成为必要，需要；强迫，迫使	C4S4
needle bed	针床	C4S4
needle felts	针状毡	C6S1
needle loop	针编弧	C4S4
needle punching	针刺	C4S5
neolithic culture	新石器文化	C3S1
Neolithic era	新石器时代	C2S1
Nobel laureate	诺贝尔奖得主	C1S2
node	（网络）节点；交点；（根或枝上的）节	C6S3
nomadic	游牧的；流浪的	C4S1
non-aesthetic purposes	非审美目的	C6S1
non-agrarian	非农业的	C4S4
non-implantable materials	非植入性材料	C6S2
non-renewable resources	不可再生资源	C5S3
non-sectarian college	［教］非教派学院	C1S2
non-volatile	（液体或油）不易挥发的	C4S2
nonwoven technology	非织造工艺	C4S5
notwithstanding	尽管，虽然	C5S3
nylon	尼龙；尼龙织品	C4S1
occupied	有人使用（或居住）；使用中	C6S3
offcut	（纸张、木材、布料等的）下脚料	C5S3
one-off	一次性的	C5S3
open lap	开口式垫纱，开口线圈	C4S4

open-ended spinning	自由端纺纱	C4S2
organic cotton	有机棉	C4S1
organic fibre	有机纤维	C4S1
orifice	孔；洞口	C4S5
ornamented pins	装饰别针	C3S1
Ottoman Empire	奥斯曼帝国	C5S1
output device	输出设备；输出装置	C6S3
packaging textiles	包装织物	C6S1
padding	填充（物）	C6S2
paisley	佩斯利（旋涡纹）图案	C3S2
pandemic	流行病	C5S2
papyrus	纸莎草	C1S1
parametre	通径；参数；变数；（结晶）半晶轴	C6S2
Parthian	帕提亚帝国（阿萨息斯王朝或安息帝国）	C5S1
particle	微粒；极少量；质点，粒子	C6S2
passementerie	珠缀；衣服的金银饰带	C3S1
patroness	（女）守护神	C2S1
pattern book	介绍（纺织品）花色的小册子	C1S2
pattern loom	提花织机	C1S1
pattern unit	图案单元；花本	C1S1
penetrate	穿过；透入；浸透；了解；进入	C6S2
peplos	（古希腊的）女式长外衣；女式大披肩	C3S1
perforate	穿孔于，在……上打眼	C4S5
permeable fabrics	透水织物	C6S1
perpendicularly	垂直地，直立地	C4S2
pervasive	遍布的；充斥各处的；弥漫的	C6S3
phase-out	逐步淘汰；分阶段撤销	C5S3
philanthropist	慈善家；乐善好施的人	C1S2

pick	纬纱；引纬	C4S3
picking	引纬	C4S3
picking/weft insertion	引纬	C4S3
pile-loop brocade	绒圈锦	C1S1
pillar/chain stitch	编链组织	C4S4
pin stripe	细条纹	C3S2
plain fabric	平针织物	C4S4
plain weave	平纹组织	C1S1
plain weave	平纹组织	C4S3
plain-woven fabric	平纹机织物	C3S2
plaster cast	石膏铸件	C6S2
plaster fabrics	石膏布	C6S2
plaster fabrics	石膏织物	C6S2
plastic surgery	整容手术	C6S2
pliability	柔韧性；可弯性	C4S1
plush	长毛绒；长绒棉	C4S1
poetry genre	诗歌流派	C2S2
poise	镇静；保持(某种姿势)；抓紧；使稳定	C5S2
polyacrylonitrile	聚丙烯腈	C4S1
polyamide	聚酰胺	C4S1
polyester	聚酯	C3S2
polyethylene	聚乙烯	C4S1
polymer	[高分子]聚合物；多聚物	C4S1
polymer chip	聚合物切片	C4S2
polymeric implants	聚合物植入物	C6S1
polypropylene	聚丙烯	C4S1
polytechnic	综合性工艺学校(大学)；理工学院	C1S2
polyurethane	聚氨酯	C4S1

pompom	小绒球；(美国啦啦队常用的)塑料丝球	C6S3
porcelain	瓷器	C2S1
power suit	力量型西装	C6S3
power supplies	电源；供电	C6S3
prefecture	县；(法、意、日等国的)地方行政区域	C1S2
premises	房产；(企业、机构的)营业场所	C1S2
presser	压板	C4S4
pressure vessel	压力容器	C6S2
pressurized	增压的；加压的	C4S3
prickle	刺；刺痛；植物的皮刺	C3S1
primary weaves	基本机织组织	C4S3
primitive looms	原始织机	C2S2
printed and dyed fabrics	印染织物	C3S2
program	项目；计划；课程(表)；培养方案	C1S2
projectile loom	片梭织机	C4S3
prominent	明显的；著名的；杰出的；广为人知的	C6S2
proof	检验；给……提供防护措施	C6S1
proofing materials	防护材料	C6S1
propel	推进；推动	C4S3
proponent	支持者；倡导者；拥护者	C6S3
protective equipment	防护装备	C6S1
protein fibre	蛋白质纤维	C4S1
prototype	原型；范例；雏形	C6S1
pucker	皱纹；皱褶	C3S2
pullover	套头毛衣；套头衫	C4S4
pulp	木浆，黏浆状物质	C4S1
puncture	击穿；穿孔；(车胎等的)刺孔	C6S3
purchasing power	购买力	C5S2

purification	净化；提纯	C6S2
purl fabric	双反面织物	C4S4
radiation	辐射；放射线	C6S1
radioactive contamination	放射性污染	C6S2
ramie	苎麻；苎麻纤维	C1S1
rapier loom	剑杆织机	C4S3
Raschel machine	拉舍尔经编机	C4S4
rash-causing plants	引起皮疹的植物	C3S1
rayon	人造丝；人造纤维	C3S2
ready-to-wear fashion	成衣时尚	C3S1
ready-to-wear fashion	成衣时装	C3S1
real-time	实时的	C6S1
redbrick university	红砖大学	C1S2
reed	钢扣	C4S3
reed split	筘隙	C4S3
reel	卷，绕上卷轴	C2S2
reeled silk	绞丝	C4S1
regalia	正式场合象征地位的服饰	C2S2
regenerated	再生的	C4S1
regenerated fibre	再生纤维	C4S1
regenerative agriculture	再生农业	C5S3
reinforcing surgical meshes	加固手术网	C6S2
relief	浮雕	C1S1
replacements surgery	置换手术	C6S2
resemble	类似；相似；相像	C6S3
reservoir	水库；储藏；蓄水池；积蓄	C6S1
resilience	弹性；弹力	C4S1
resiliency	弹性；弹力	C4S1

resilient	弹回的，有弹力的；能复原的；可迅速恢复的	C5S2
resist	防染布料	C1S2
resist dying	防染；扎染；夹缬	C1S1
resistance	电阻；抵抗；阻力；抗力	C6S3
resource-intensive	资源密集型	C5S2
restorative	有助于复原的	C5S3
resurgence	复苏，复活；中断之后的继续	C4S4
retail	零售；（以某种价格）零售	C5S2
retail market	零售市场	C5S2
reverence	崇敬	C2S1
reverse loop stitch	反面线圈组织	C4S4
revolution	旋转，绕转	C1S1
rib	罗纹	C4S4
rib fabric	罗纹织物	C4S4
rigid ballistic panels	刚性弹道板	C6S2
ring spinning	环锭纺	C4S2
ring spun yarn	环锭纺纱	C4S2
ritual	仪式，典礼	C2S1
rivet	针舌销；铆钉	C4S4
robust	强健的；强壮的；富有活力的	C5S3
rotor spinning	转杯纺纱	C4S2
roving frame	粗纱机	C4S2
royal charter	（英国）皇家特许状	C1S2
rush	灯芯草	C1S1
sash	腰带；肩带	C3S1
sateen weave	纬面缎纹	C4S3
satin	缎子	C2S1
satin weave	经面缎纹	C4S3

schematic diagram	原理图，示意图	C4S5
schools of embroidery	刺绣流派	C2S2
scroll	长卷纸，卷轴	C2S2
scrubs	外科手术服	C3S1
sea island cotton	海岛棉	C4S1
sea routes	海上航线	C2S1
seal	印章，标志	C2S1
sedentary	坐着的；(指人)不爱活动的	C4S1
semi-synthetic fabric	半合成织物	C3S2
sensor	(探测光、热、压力等的)传感器	C6S3
sensual	感官的；肉体的	C6S2
serge	毛哔叽	C4S3
sericulture	养蚕	C2S1
sewerage	排水设备；污水	C6S1
sewn leather	缝制皮革	C3S1
shawl	披肩；围巾	C1S2
shearing	剪羊毛，剪取的羊毛	C4S1
shed	梭口	C1S1
shedding	开口	C4S3
sheeting	粗平布，被单料子	C4S3
shell fabric	面料	C4S3
shirting	细平布，衬衫衣料	C4S3
shroud	寿衣；裹尸布	C1S1
Shu Brocade	蜀锦	C2S2
Shu Embroidery	蜀绣	C2S2
shuttle	梭；梭子	C2S2
shuttle loom	有梭织机	C4S3
shuttleless loom	无梭织机	C4S3

side limb/pillar	圈柱	C4S4
silhouette	轮廓，剪影	C3S2
silicon carbonization fibre	碳化硅纤维	C4S1
silk cloth	绸布	C2S1
silk fabric	真丝织物	C3S2
silk road	丝绸之路	C5S1
silk-knit goods	针织品	C2S2
simple tunic	简单束腰外衣	C3S1
single faced fabric	单面织物	C4S4
single warp thread	单经线	C3S2
single-spool hand wheel	单轴手轮	C2S1
singular	非凡的；突出的；奇异的	C1S1
sinker loop	沉降弧	C4S4
sinker wheel	沉降轮	C4S4
sinuous	弯曲有致的；蜿蜒的	C1S1
sinuous curves	优美的曲线	C1S1
sizing	上浆	C4S3
slashing	上浆	C4S3
slider	（复合针）针芯	C4S4
slight single rib knit	轻薄单罗纹针织	C3S2
sliver	条子，梳条	C4S2
small-scale peasant economy	小农经济	C2S2
smart textiles	智能纺织	C5S2
smocking	装饰用衣褶	C3S2
snippet	（消息、知识等）的片断；零星的话	C6S3
soft drapery chiffon skirt	软帘雪纺半身裙	C3S2
soft laced shoes	软花边鞋	C3S1
solar panel	太阳能电池板	C6S3

soluble	[化]可溶的	C4S2
solution	溶液；溶解	C4S2
solution spinning	溶液纺丝	C4S2
solvent	[化]溶剂，溶媒	C4S2
Song Brocade	宋锦	C2S2
soothing colour	舒缓的色彩	C3S2
sort	整理；把……分类	C5S3
sound-proofing element	隔音元素	C6S1
soybean fibre	大豆纤维	C4S1
spandex	氨纶，弹性纤维	C3S2
specify	详细说明；指定；阐述	C6S1
speculate	推测；投机；思索；推断	C5S1
spin	纺纱；纺丝	C4S1
spindle	轴；纺锤；纱锭	C1S1
spindle wheel, spinning wheel	纺车	C1S1
spindle whorl	锭盘	C1S1
spinneret	喷丝头；喷丝板	C4S1
spinning frame	纺纱机	C2S1
spin-off	周边产品；搭车产品	C6S3
splinter	碎片；微小的东西	C3S1
sport	夸示；得意地穿戴；故意显示	C6S3
sports clothing	运动服装	C6S1
spring power	弹簧动力	C4S3
spun yarn/staple yarn	短纤纱	C4S2
spunbonded nonwovens	纺粘无纺布	C4S5
stainless-steel	不锈钢	C6S3
staple	主要产品，支柱产品；纤维，短纤维	C2S1

staple fibre	短纤维	C4S1
startup	初创公司，新成立的公司	C6S3
stationary	不动的；静止的；固定的	C6S3
statistic monitoring	数据监控	C6S1
steer	驾驶汽车；操舵	C6S2
stem	针杆	C4S4
stencil	（印文字或图案用的）模板；（油印）蜡纸	C1S2
sterilize	消毒；使无菌	C4S2
stitch	针法；针脚	C2S1
stitch bonder	缝编机	C4S5
stitched bonding	缝编黏合	C4S5
stocking frame	织袜机	C4S4
stola	女士及踝长外衣	C3S1
strand	（绳子的）股，绞	C4S2
strapping	带子	C4S4
stretchable	有弹性的	C4S4
stroke	抚摸；轻触；[纺]弄伸皱褶	C6S3
strongest visual impact	最强视觉冲击力	C3S2
sturdy	坚固的，耐用的	C4S3
Su Embroidery	苏绣	C2S2
sunscreen materials	防晒材料	C6S1
suppress	镇压；隐瞒；压制；止住；禁止	C6S1
surgical clothing	外科手术服	C6S1
surplus goods	剩余商品	C2S2
survivable limits	可存活极限	C6S2
susceptible	易受影响的；易受感染的	C4S1
sustainability	持续性，能维持性	C4S1
suture	缝合，缝伤口；缝线	C4S2

sweater	毛衣，运动衫	C4S4
symmetrical organization	对称组织	C1S1
synthesis	<化>合成	C4S1
synthetic fibres	合成纤维	C3S1
synthetic fibre	合成纤维	C4S1
synthetic materials	合成材料	C6S2
synthetic paints	合成油漆	C3S1
take-make-dispose model	取舍模式	C5S3
taking up	牵拉	C4S3
Tang embroidery art	唐绣艺术	C2S2
Tang silks	唐代丝绸	C2S1
tangible	清晰明确的	C5S2
tapestry	挂毯；织锦；壁毯；绣帷	C1S1
technical back	工艺反面	C4S4
technical face	工艺正面	C4S4
technical textiles	技术纺织品	C6S1
temperature-resistant clothing	耐温服装	C6S2
template	模板	C2S2
tenacity	韧性；柔韧性	C4S1
tex	特克斯	C4S2
textile architecture	纺织建筑学	C6S1
textile fibre	纺织纤维	C4S1
textro-polymer	高分子电子纺织材料	C6S3
texture	质地；手感	C2S1
the central plains	中原地区	C2S2
the craft of embroidery	刺绣工艺	C2S2
the Silk Road	丝绸之路	C2S1

the spinning jenny	珍妮纺纱机	C2S1
The Travels of Marco Polo	《马可·波罗游记》	C5S1
thermal	热的，保热的；温热的	C4S5
thermal and acoustic protection	放热防噪音保护	C6S2
thermal bonding	热黏合	C4S5
thermal conductivity	导热系数	C6S2
thermal insulation	隔热	C6S1
thermal insulation properties	隔热性能	C3S1
thermochromic ink	热敏油墨	C6S3
thermoplastic	热塑性塑料	C4S5
thick woolen long dresses	粗呢长连衣裙	C3S1
thorn	刺；荆棘	C3S1
thoroughfare	大道，通路	C5S1
three-spool, pedal-driven cotton-spinning machine	脚踏式三轴棉纺机	C2S1
tie-dyeing	扎染	C2S1
tip	针尖	C4S4
to conduct electricity	导电	C6S3
toga	(古罗马的)宽外袍	C3S1
tongue	(复合针)针芯	C4S4
toxic	有毒的；中毒的；有害的；致命的；恶毒的	C6S2
traceability	可追溯性	C5S2
transcontinental	横贯大陆的	C2S1
transcontinental trade	跨洲贸易	C2S1
transformative	革命性的；转折的	C5S2
transverse	横向的；横断的	C4S4
traverse	穿过	C2S1

treadle	（尤指旧时驱动机器的）踏板	C1S1
treadle loom	踏式织机；蹑机	C1S1
triacetate	三醋酸酯，三乙酸酯	C4S1
Tricot machine	特利考经编机	C4S4
tricot stitch	经平组织	C4S4
trustee	受托人；保管人	C1S2
turbulent	激流的，湍流的	C4S5
turnover	营业额，成交量	C1S1
tussah	柞蚕丝；柞蚕，野蚕丝	C4S1
twill	斜纹布	C1S1
twill damask	斜纹锦缎	C1S1
twill weave	斜纹组织	C4S3
twining foliage	缠枝纹	C1S1
twist	捻；加捻	C4S2
twist level	捻度	C4S2
two-ply	双股的	C4S1
ultraviolet-resistant clothing	防紫外线服装	C6S2
unbleached	原色的，未经漂白的	C2S2
undergarment	内衣	C2S1
underlap	延展线	C4S4
uniform	均匀的；一致的	C4S2
uniform rubber mass	均匀橡胶质量	C6S2
uniformity	均匀性	C4S1
unravel	脱散	C4S4
upholstery	家具装饰品，坐垫和用来覆盖的织物	C1S1
upholstery fabric	室内装饰面料	C1S1
upland cotton	陆地棉	C4S1

user comfort and safety	用户舒适度和安全感	C6S1
UV radiation	紫外线；紫外辐射	C3S1
value chain	价值链	C5S3
valve repair	（心脏）瓣膜修复	C6S2
vegetable fibre, plant fibre	植物纤维	C1S1
velvet	丝绒；天鹅绒；立绒	C4S2
vertical loom, upright loom	立式织机；竖机	C1S1
vibration	振动；动摇；不安	C6S2
viceroy	（旧时受君主委派管治殖民地的）总督	C1S2
vinyl	乙烯基（化学）	C3S2
viscose	纤维胶；人造丝	C3S2
viscosity	黏稠；黏质；黏性	C4S5
visual	视觉的；视力的；看得见的；形象的	C6S1
Vogue magazine	《时尚》杂志	C4S4
volatile	（液体或油）易挥发的	C4S2
voltage gradient	电压梯度；电压陡度	C6S3
voluminous	大量的	C3S1
voluminous dresses	宽大的连衣裙	C3S1
voluminous skirt	宽松半身裙	C3S1
wale	纵行	C4S4
wall-hanging	墙帷；挂饰	C6S3
warm colours	暖色调	C3S2
warp	经纱	C1S1
warp beam	经轴	C1S1
warp face satin	经面缎纹	C4S3
warp knitting	经编针织	C4S4
warp patterning	经丝显花	C1S1
warping	整经	C4S3

Warring States Period	战国时期	C2S2
water jet loom	喷水织机	C4S3
weather resistance	耐候性	C6S1
weaving loom	织机	C4S3
weaving loom	织机	C4S3
weaving techniques	织造技术	C2S1
weaving technology	织造工艺	C4S3
weft	纬纱	C1S1
weft face (filling face) satin	纬面缎纹	C4S3
weft knitting	纬编针织	C4S4
weft patterning	纬丝显花	C1S1
weft thread	纬线	C3S2
weight based system/fixed length system	定长制	C4S2
wet spinning	湿法纺丝	C4S2
wild silk	野蚕丝；柞蚕丝	C4S1
winding	络筒	C4S3
with a view to	为了；着眼于	C6S3
wool blend	羊毛混纺	C3S2
wool textiles	羊毛纺织品	C2S1
woven fibres	机织纤维	C3S1
woven silk brocade	丝织锦缎	C3S2
woven technology	机织工艺	C4S5
woven textile	机织纺织品	C4S1
Xiang Embroidery	湘绣	C2S2
yarn	纱线	C1S1
yarn count	纱线支数	C4S2

yarn feeding	喂纱	C4S4
Yue Embroidery	粤绣	C2S2
Yuezhi	大宛	C5S1
Zhuang Brocade	壮锦	C2S2
z-twist crepe yarns	Z捻绉纱	C3S2

References

参考书目

[1] Cheng, Weiji. *History of Textile Technology of Ancient China* [M]. New York: Science Press New York, Ltd. and Science Press, 1992.

[2] Chunguang Ren, Xiaoming Yang. *On Textile Poetry in Textile Social History in Qin and Han Dynasties* [J]. Canadian Center of Science and Education, 2020.

[3] Harris, Jennifer. *5000 Years of Textiles* [M]. Singapore: The British Museum Press, 2010.

[4] LinHanda, Cao Yuzhang. *Tales from 5000 Years of Chinese History* [M]. New York: Better Link Press, 2010.

[5] Wang Jian, Fang Xiaoyan. *An Illustrated Brief History of China* [M]. New York: Better Link Press, 2017.

[6] 陈维稷. 中国纺织科学技术史(古代部分) [M]. 北京: 科学出版社, 1984.

[7] 陈毅平, 秦学信. 大学英语文化翻译教程 [M]. 北京: 外语教学与研究出版社, 2014.

[8] 高秀明. 服装时尚行业英语 [M]. 上海: 东华大学出版社, 2014.

[9] 郭红梅, 崔磊, 杨晶. 中国文化通识基础翻译教程 [M]. 北京: 冶金工业出版社, 2020.

[10] 华梅. 中国服饰 [M]. 北京: 五洲传播出版社, 2004.

[11] 刘莹莹, 李岚, 何丹霞. 用英语介绍中国 [M]. 北京: 化学工业出版社, 2014.

[12] 佟昀. 纺织应用英语 [M]. 北京: 中国纺织出版社, 2014.

[13] 赵承泽. 中国科学技术史·纺织卷 [M]. 北京: 科学出版社, 2002.

[14] 钟智丽. 中国纺织业与纺织先进技术(英文版) [M]. 北京: 中国纺织出版社, 2020.

[15] 庄三生. 服装专业英语 [M]. 上海: 东华大学出版社, 2015.

[16] 卓乃坚. 纺织英语(第三版) [M]. 上海: 东华大学出版社, 2017.

参考网站

[1] https://www.economist.com/technology-quarterly/2005/12/10/threads-that-think.

[2] https://en.wikipedia.org/wiki/History_of_clothing_and_textiles.

[3] https://en.wikipedia.org/wiki/Textile.

References

[4] https://www.britannica.com/biography/Ada-Lovelace.

[5] https://writings.stephenwolfram.com/2015/12/untangling-the-tale-of-ada-lovelace.

[6] https://www.leeds.ac.uk/info/5000/about/133/heritage.

[7] https://www.manchester.ac.uk/discover/history-heritage/history.

[8] https://www.ncsu.edu/about.

[9] https://en.wikipedia.org/wiki/RWTH_Aachen_University.

[10] https://en.wikipedia.org/wiki/Moscow_State_Textile_University.

[11] https://en.wikipedia.org/wiki/IIT_Delhi.

[12] https://en.wikipedia.org/wiki/Shinshu_University.

[13] http://english.wtu.edu.cn/About/About_WTU.htm.

[14] https://www.sciencedirect.com/topics/engineering/textile-science.

[15] http://textileconservation.academicblogs.co.uk/about-us.

[16] https://www.sohu.com/a/455170790_116132.

[17] https://www.sciencedirect.com/topics/engineering/electronic-textile.

[18] https://en.wikipedia.org/wiki/Huang_Daopo.

[19] https://archive.shine.cn/sunday/now-and-then/%E9%BB%84%E9%81%93%E5%A9%86-Huang-Daopo-circa-12451330-Grannys-great-innovations/shdaily.shtml.

[20] https://www.marketplace.org/2016/11/25/how-woman-revolutionized-her-small-towns-economy.

[21] https://fashion-history.lovetoknow.com/fabrics-fibers/chinese-textiles.

[22] https://en.wikipedia.org/wiki/Textile.

[23] https://en.wikipedia.org/wiki/History_of_silk.

[24] https://en.wikipedia.org/wiki/Silk.

[25] https://study.com/academy/lesson/ancient-chinese-textiles.html.

[26] http://www.textileasart.com/weaving.htm#chinese.

[27] https://www.travelchinaguide.com/intro/arts/embroidery.htm.

[28] http://countrystudies.us/china/73.htm.

[29] http://silkroadfoundation.org/artl/silkhistory.shtml.

[30] http://www.silk-road.com/artl/silkhistory.shtml#:~:text=%20Silk%20history%20%20201%20Early%20silk%20in,of%20China%20and%20those%20very%20close...%20More%20.

[31] http://www.historyofclothing.com/textile-history/history-of-cotton.

[32] https://www.esl-lab.com/basic-english/clothing-and-fashion.

[33] https://www.dailyesl.com/at-home/clothing-fashion.

[34] https://www.textileschool.com/amp/4639/origin-of-clothing.

[35] http://www.historyofclothing.com.

[36] http://www.historyofclothing.com/textile-history.

[37] http://www.historyofclothing.com/making-clothing.

[38] https://www.britannica.com/topic/textile.
[39] https://mask.xmtextiles.com/what-is-spunbond-meltblown-fabric-and-why-it-is-in-face-mask.
[40] https://textilelearner.net.
[41] https://fr.wikipedia.org/wiki/Textile.
[42] https://fr.wikipedia.org/wiki/Fibre_textile.
[43] https://www.textileschool.com.
[44] https://www.britannica.com/topic/natural-fiber.
[45] https://www.textileschool.com/486/synthetic-fibers-manmade-artificial-fibers.
[46] https://www.britannica.com/technology/weaving.
[47] https://cn.bing.com/images/search? q = knitting&qpvt = knitting&form = IGRE&first = 1&tsc = ImageBasicHover.
[48] https://cn.bing.com/images/search? q = weaving&go = Search&qs = ds&form = QBIR&first = 1&tsc = ImageBasicHover.
[49] https://www.thecreativefolk.com/difference-warp-and-weft-knitting.
[50] https://www.textileblog.com/nonwoven-fabric-manufacturing-techniques/#:~:text = Nonwoven%20manufacturing%20techniques%20is%20basically%20a%20continuous%20process, two%20basic%20steps%2C%20web%20formation%20followed%20by%20bonding.
[51] https://www.history.com/topics/ancient-middle-east/silk-road.
[52] http://www.silk-road.com.
[53] https://www.travelchinaguide.com/silk-road.
[54] https://www.inkworldmagazine.com/contents/view_breaking-news.
[55] http://www.globaltextiles.com.
[56] http://www.ittaindia.org/? q = abouttechnicaltextile.
[57] http://www.innovationintextiles.com.
[58] https://www.ellenmacarthurfoundation.org.